Use R!

Series Editors:
Robert Gentleman   Kurt Hornik   Giovanni Parmigiani

Use R!

Albert: Bayesian Computation with R

Bivand/Pebesma/Gómez-Rubio: Applied Spatial Data Analysis with R

Claude: Morphometrics with R

Cook: Interactive and Dynamic Graphics for Data Analysis

Hahne/Huber/Gentleman/Falcon: Bioconductor Case Studies

Nason: Wavelet Methods in Statistics with R

Paradis: Analysis of Phylogenetics and Evolution with R

Peng/Dominici: Statistical Methods for Environmental Epidemiology with R

Pfaff: Analysis of Integrated and Cointegrated Time Series with R

Sarkar: Lattice: Multivariate Data Visualization with R

Spector: Data Manipulation with R

Bernhard Pfaff

Analysis of Integrated and Cointegrated Time Series with R

 Springer

Dr. Bernhard Pfaff
61476 Kronberg im Taunus
Germany

Series Editors:
Robert Gentleman
Program in Computational Biology
Division of Public Health Sciences
Fred Hutchinson Cancer Research Center
1100 Fairview Ave. N, M2-B876
Seattle, Washington 981029-1024
USA

Kurt Hornik
Department für Statistik und Mathematik
Wirtschaftsuniversität Wien Augasse 2-6
A-1090 Wien
Austria

Giovanni Parmigiani
The Sidney Kimmel Comprehensive Cancer Center at Johns Hopkins University
550 North Broadway
Baltimore, MD 21205-2011
USA

ISBN 978-0-387-75966-1   e-ISBN 978-0-387-75967-8

Library of Congress Control Number: 2008930126

Printed on acid-free paper

9 8 7 6 5 4 3 2 1

springer.com

To my parents

Preface to the Second Edition

A little more than two years have passed since the first edition. During this time, R has gained further ground in the domain of econometrics. This is witnessed by the 2006 useR! conference in Vienna, where many sessions were devoted entirely to econometric topics, as well as the Rmetrics workshop at Meielisalp 2007. A forthcoming special issue of the *Journal of Statistical Software* will be devoted entirely to econometric methods that have been implemented within R. Furthermore, numerous new packages have been contributed to CRAN and existing ones have been improved; a total of more than 1200 are now available. To keep up with these pleasant changes, it is therefore necessary not only to adjust the R code examples from the first edition but also to enlarge the book's content with new topics.

However, the book's skeleton and intention stays unchanged, given the positive feedback received from instructors and users alike. Compared with the first edition, vector autoregressive (VARs) models and structural vector autoregressive (SVARs) models have been included in an entire new chapter in the first part of the book. The theoretical underpinnings, definitions, and motivation of VAR and SVAR models are outlined, and the various methods that are applied to these kinds of models are illustrated by artificial data sets. In particular, it is shown how swiftly different estimation principles, inference, diagnostic testing, impulse response analysis, forecast error variance decomposition, and forecasting can be conducted with R. Thereby the gap to vector error-correction models (VECMs) and structural vector error-correction (SVEC) models is bridged. The former models are now introduced more thoroughly in the last chapter of the first part, and an encompassing analysis in the context of VEC/SVEC modeling is presented in the book's last chapter. As was the case for the first edition, all R code examples presented can be downloaded from http://www.pfaffikus.de.

As with the first edition, I would like to thank the R Core Team for providing such a superb piece of software to the public and to the numerous package authors who have enriched this software environment. I would further like to express my gratitude to the anonymous referees who have given good pointers for improving this second edition. Of course, all remaining errors are mine. Last but not least, I would like to thank my editor, John Kimmel, for his continuous encouragement and support.

Kronberg im Taunus   *Bernhard Pfaff*
March 2008

Preface

This book's title is the synthesis of two influential and outstanding entities. To quote David Hendry in the Nobel Memorial Prize lecture for Clive W. J. Granger, "[the] modeling of non-stationary macroeconomic time series [...] has now become the dominant paradigm in empirical macroeconomic research" (Hendry [2004]). Hence, a thorough command of integration and cointegration analysis is a must for the applied econometrician. On the other side is the open-source statistical programming environment R. Since the mid-1990s, it has grown steadily out of infancy and can now be considered mature, flexible, and powerful software with more than 600 contributed packages. However, it is fair to say that R has not yet received the attention among econometricians it deserves. This book tries to bridge this gap by showing how easily the methods and tools encountered in integration and cointegration analysis are implemented in R.

This book addresses senior undergraduate and graduate students and practitioners alike. Although the book's content is not a pure theoretical exposition of integration and cointegration analysis, it is particularly suited as an accompanying text in applied computer laboratory classes. Where possible, the data sets of the original articles have been used in the examples such that the reader can work through them step by step and thereby replicate the results. Exercises are included after each chapter. These exercises are written with the aim of fostering the reader's command of R and applying the previously presented tests and methods. It is assumed that the reader has already gained some experience with R by working through the relevant chapters in Dalgaard [2002] and Venables and Ripley [2002] as well as the manual "An Introduction to R."

This book is divided into three parts. In the first part, theoretical concepts of time series analysis, unit root processes, and cointegration are presented. Although the book's aim is not a thorough theoretical exposition of these methods, this first part serves as a unifying introduction to the notation used and as a brief refresher of the theoretical underpinnings of the practical examples in the later chapters. The focus of the second part is the testing of the unit root hypothesis. The common testing procedure of the augmented

Dickey-Fuller test for detecting the order of integration is considered first. In the later sections, other unit root tests encountered widely in applied econometrics, such as the Phillips-Perron, Elliott-Rothenberg-Stock, Kwiatkowski-Phillips-Schmidt-Shin, and Schmidt-Phillips tests, are presented, as well as the case of seasonal unit roots and processes that are contaminated by structural shifts. The topic of the third and last part is cointegration. As an introduction, the two-step method of Engle and Granger and the method proposed by Phillips and Ouliaris are discussed before finally Johansen's method is presented. The book ends with an exposition of vector error-correction models that are affected by a one-time structural shift.

At this point, I would like to express my gratitude to the R Core Team for making this software available to the public and to the numerous package authors who have enriched this software environment. The anonymous referees are owed a special thanks for the suggestions made. Of course, all remaining errors are mine. Last but not least, I would like to thank my editor, John Kimmel, for his continuous encouragement and support.

Kronberg im Taunus   *Bernhard Pfaff*
September 2005

Contents

Preface to the Second Edition vii

Preface ix

List of Tables xv

List of Figures xvii

List of R Code xix

Part I Theoretical Concepts

1   Univariate Analysis of Stationary Time Series 3
    1.1   Characteristics of Time Series 3
    1.2   AR(p) Time Series Process 6
    1.3   MA(q) Time Series Process 10
    1.4   ARMA(p, q) Time Series Process 14
    Summary 20
    Exercises 21

2   Multivariate Analysis of Stationary Time Series 23
    2.1   Overview 23
    2.2   Vector Autoregressive Models 23
          2.2.1   Specification, Assumptions, and Estimation 23
          2.2.2   Diagnostic Tests 28
          2.2.3   Causality Analysis 34
          2.2.4   Forecasting 36
          2.2.5   Impulse Response Functions 37
          2.2.6   Forecast Error Variance Decomposition 41
    2.3   Structural Vector Autoregressive Models 43
          2.3.1   Specification and Assumptions 43

2.3.2   Estimation 44
2.3.3   Impulse Response Functions 47
2.3.4   Forecast Error Variance Decomposition 48
Summary 49
Exercises 50

3   Non-stationary Time Series 53
3.1   Trend- versus Difference-Stationary Series 53
3.2   Unit Root Processes 55
3.3   Long-Memory Processes 62
Summary 70
Exercises 71

4   Cointegration 73
4.1   Spurious Regression 73
4.2   Concept of Cointegration and Error-Correction Models 75
4.3   Systems of Cointegrated Variables 78
Summary 86
Exercises 86

Part II Unit Root Tests

5   Testing for the Order of Integration 91
5.1   Dickey-Fuller Test 91
5.2   Phillips-Perron Test 94
5.3   Elliott-Rothenberg-Stock Test 98
5.4   Schmidt-Phillips Test 100
5.5   Kwiatkowski-Phillips-Schmidt-Shin Test 103
Summary 104
Exercises 105

6   Further Considerations 107
6.1   Stable Autoregressive Processes with Structural Breaks 107
6.2   Seasonal Unit Roots 112
Summary 118
Exercises 118

Part III Cointegration

7   Single-Equation Methods 121
7.1   Engle-Granger Two-Step Procedure..........................121
7.2   Phillips-Ouliaris Method 123
Summary 126
Exercises 127

8   Multiple-Equation Methods 129
  8.1   The Vector Error-Correction Model 129
     8.1.1   Specification and Assumptions 129
     8.1.2   Determining the Cointegration Rank................. 130
     8.1.3   Testing for Weak Exogenity 134
     8.1.4   Testing Restrictions on β 136
  8.2   VECM and Structural Shift............................. 143
  8.3   The Structural Vector Error-Correction Model 145
  Summary 158
  Exercises 158

9   Appendix... 161
  9.1   Time Series Data.. 161
  9.2   Technicalities 162
  9.3   CRAN Packages Used 163

10 Abbreviations, Nomenclature, and Symbols................ 165

References... 169

Name Index 177

Function Index 181

Subject Index 185

List of Tables

2.1 VAR result for y_1 26
2.2 VAR result for y_2 27
2.3 Eigenvalues of the companion form 27
2.4 Diagnostic tests of VAR(2) 33
2.5 Causality tests 36
2.6 SVAR A-model: estimated coefficients........................ 45
2.7 SVAR B-model: estimated coefficients........................ 47
2.8 SVAR B-model: FEVD for y_2 49
2.9 Overview of package **vars** 50

3.1 Results of Geweke and Porter-Hudak method 70

4.1 Results of spurious regression 75
4.2 Results of Engle-Granger procedure with generated data 78
4.3 Cointegration rank: Maximum eigenvalue statistic 83
4.4 Cointegration vectors 83
4.5 Normalized cointegration vectors 84
4.6 VECM with $r = 2$ 85

5.1 ADF test: Regression for consumption with constant and trend. 92
5.2 ADF test: τ_3, ϕ_2, and ϕ_3 tests 93
5.3 ADF test: Regression for consumption with constant only 94
5.4 ADF test: τ_2 and ϕ_1 tests 94
5.5 ADF test: Regression for testing $I(2)$ 95
5.6 PP test: Regression for consumption with constant and trend .. 97
5.7 PP test: $Z(t_{\tilde{\alpha}})$, $Z(t_{\tilde{\mu}})$, and $Z(t_{\tilde{\beta}})$ tests 97
5.8 PP test: Regression for consumption with constant only 97
5.9 PP test: $Z(t_{\hat{\alpha}})$ and $Z(t_{\hat{\mu}})$ tests 97
5.10 PP test: Regression for testing $I(2)$ 98
5.11 ERS tests: P_T^{τ} and DF-GLS$^{\tau}$ for real GNP of the United States 100

5.12 ERS tests: P_T^τ and DF-GLS$^\tau$ for testing $I(2)$ 100

5.13 SP test: Result of level regression with polynomial of order two. 102

5.14 SP tests: $\tilde{\tau}$ and $\tilde{\rho}$ for nominal GNP of the United States 102

5.15 KPSS tests for interest rates and nominal wages of the United States 104

6.1 Zivot-Andrews: Test regression for nominal wages 111

6.2 Zivot-Andrews: Test regression for real wages 112

6.3 Zivot-Andrews: Test statistics for real and nominal wages 112

6.4 Cycles implied by the roots of the seasonal difference operator . 114

6.5 HEGY test: Real disposable income in the United Kingdom ... 118

7.1 Engle-Granger: Cointegration test 123

7.2 Engle-Granger: ECM for the consumption function 124

7.3 Engle-Granger: ECM for the income function 124

7.4 Phillips-Ouliaris: Cointegration test 126

8.1 Cointegration rank: Maximal eigenvalue statistic 132

8.2 Cointegration rank: Trace statistic 133

8.3 $\mathcal{H}_1$ model: Eigenvectors 133

8.4 $\mathcal{H}_1$ model: Weights 133

8.5 $\mathcal{H}_1(2)$ model: Coefficient matrix of the lagged variables in levels, $\hat{\boldsymbol{\Pi}}$ 134

8.6 $\mathcal{H}_4$ model: Testing for weak exogeneity 137

8.7 $\mathcal{H}_3$ model: Restriction in all cointegration relations 138

8.8 $\mathcal{H}_5$ model: Partly known cointegration relations 140

8.9 $\mathcal{H}_6$ model: Restrictions on r_1 cointegration relations 142

8.10 Money demand function for Denmark: Maximal eigenvalue statistic, non-adjusted data 145

8.11 Money demand function for Denmark: Trace statistic, allowing for structural shift 145

8.12 ADF tests for Canadian data 149

8.13 Canada VAR: Lag-order selection 149

8.14 Diagnostic tests for VAR(p) specifications for Canadian data... 151

8.15 Johansen cointegration tests for Canadian system 153

8.16 Cointegration vector and loading parameters 154

8.17 Estimated coefficients of the contemporaneous impact matrix .. 156

8.18 Estimated coefficients of the long-run impact matrix 156

8.19 Forecast error variance decomposition of Canadian unemployment 158

9.1 Overview of packages used 163

List of Figures

1.1 U.S. GNP (real, logarithm) 4
1.2 U.S. unemployment rate (in percent)........................ 5
1.3 Time series plot of AR(1)-process, $\phi = 0.9$ 8
1.4 Time series plot of AR(2)-process, $\phi_1 = 0.6$ and $\phi_2 = -0.28$ ... 11
1.5 Unit circle and roots of stable AR(2)-process, $\phi_1 = 0.6$ and
    $\phi_2 = -0.28$ 13
1.6 Time series plot of MA(1)-process, $\theta = 0.8$................... 14
1.7 Time series plot, ACF, and PACF of U.S. unemployment rate.. 18
1.8 Time series plot, ACF, and Ljung-Box values of residuals
    [ARMA(1, 1)]... 19
1.9 Actual and forecasted values of the U.S. unemployment rate ... 21

2.1 Time series plot of the simulated VAR(2)-process.............. 27
2.2 Diagnostic residual plot for y_1 of VAR(2)-process 32
2.3 Diagnostic residual plot for y_2 of VAR(2)-process 33
2.4 OLS-CUSUM test for y_1 of VAR(2)-process.................. 34
2.5 Fluctuation test for y_2 of VAR(2)-process 35
2.6 Forecasting y_1 of VAR(2)-process.......................... 38
2.7 Fanchart of y_2 of VAR(2)-process.......................... 39
2.8 Impulse responses of y_1 to y_2 41
2.9 Impulse responses of y_2 to y_1 42
2.10 FEVD for VAR(2)-process................................ 43
2.11 IRA from y_1 to y_2 of SVAR A-model 49

3.1 Time series plot of deterministic and stochastic trends 57
3.2 Plot of a random walk and a stable AR(1)-process, $\phi = 0.99$ ... 58
3.3 Testing sequence for unit roots 63
3.4 Graphical display: ARIMA versus ARFIMA 67

4.1 Three $I(1)$-series with two cointegration relations............. 84

5.1   ADF test: Diagram of fit and residual diagnostics 93
5.2   Nominal GNP of the United States 102

6.1   Time series plot of random walk with drift and structural break 109
6.2   Zivot-Andrews test statistic for nominal wages 113
6.3   Zivot-Andrews test statistic for real wages 114

8.1   Graphical display of purchasing power parity and uncovered
      interest rate parity for the United Kingdom 130
8.2   Graphical display of the first two cointegration relations 134
8.3   Canadian data set: Time series plots 148
8.4   OLS-CUSUM test of VAR(3) 151
8.5   OLS-CUSUM test of VAR(2) 152
8.6   OLS-CUSUM test of VAR(1) 153
8.7   Responses of unemployment to economic shocks with a 95%
      bootstrap confidence interval 157

List of R Code

1.1 Simulation of AR(1)-process with $\phi = 0.9$ 7
1.2 Estimation of AR(2)-process with $\phi_1 = 0.6$ and $\phi_2 = -0.28$ 12
1.3 Box-Jenkins: U.S. unemployment rate 17
1.4 Box-Jenkins: Predictions of the U.S. unemployment rate 20

2.1 Simulation of VAR(2)-process 28
2.2 Diagnostic tests of VAR(2)-process 30
2.3 Empirical fluctuation processes 34
2.4 Causality analysis of VAR(2)-process 36
2.5 Forecasts of VAR-process 38
2.6 IRA of VAR-process 40
2.7 FEVD of VAR-process 42
2.8 SVAR: A-model 46
2.9 SVAR: B-model 47
2.10 SVAR: Impulse response analysis 48
2.11 SVAR: Forecast error variance decomposition 48

3.1 Stochastic and deterministic trends 56
3.2 ARMA versus ARFIMA model 66
3.3 R/S statistic 68
3.4 Geweke and Porter-Hudak method 69

4.1 Spurious regression 74
4.2 Engle-Granger procedure with generated data 77
4.3 Johansen method with artificially generated data 83
4.4 VECM as VAR in levels 85

5.1 ADF test: Integration order for consumption in the United
    Kingdom 92
5.2 PP test: Integration order for consumption in the United
    Kingdom 96

5.3   ERS tests: Integration order for real GNP in the United States . 99

5.4   SP test: Integration order for nominal GNP of the United States 101

5.5   KPSS test: Integration order for interest rates and nominal
      wages in the United States 104

6.1   Random walk with drift and structural break 108

6.2   Unit roots and structural break: Zivot-Andrews test 111

6.3   HEGY test for seasonal unit roots 117

7.1   Engle-Granger: Long-run relationship of consumption, income,
      and wealth in the United Kingdom 122

7.2   Engle-Granger: ECM for consumption and income of the
      United Kingdom 123

7.3   Phillips-Ouliaris: Long-run relationship of consumption,
      income, and wealth in the United Kingdom 126

8.1   Johansen-Juselius: Unrestricted cointegration 131

8.2 $\mathcal{H}_1$ model: Transformations and cointegration relations 133

8.3 $\mathcal{H}_4$ model: Testing for weak exogeneity 136

8.4 $\mathcal{H}_3$ model: Testing for restrictions in all cointegration relations . 138

8.5 $\mathcal{H}_3$ model: Testing for partly known cointegration relations 140

8.6 $\mathcal{H}_6$ model: Testing of restrictions on r_1 cointegration relations . . 142

8.7 $\mathcal{H}_1$ model: Inference on cointegration rank for Danish money
      demand function allowing for structural shift 144

8.8   Canadian data set: Preliminary analysis 147

8.9   Canadian data set: ADF-test regressions 148

8.10 Canada VAR: Lag-order selection 149

8.11 Diagnostic tests for VAR(p) specifications for Canadian data . . . 150

8.12 Johansen cointegration tests for Canadian system 152

8.13 VECM with $r = 1$ and normalization with respect to real wages 154

8.14 Estimation of SVEC with bootstrapped t statistics 155

8.15 SVEC: Overidentification test 156

8.16 SVEC: Impulse response analysis 157

8.17 Forecast error variance decomposition of Canadian
      unemployment 158

9.1   Time series objects of class `ts` 161

Part I

Theoretical Concepts

1

Univariate Analysis of Stationary Time Series

Although this book has integration and cointegration analysis as its theme, it is nevertheless a necessity to first introduce some concepts of stochastic processes as well as the stationary ARMA model class. Having paved this route, the next steps (i.e., the introduction of non-stationary, unit root, and long-memory processes) will follow in Chapter 3.

1.1 Characteristics of Time Series

A discrete *time series*[1] is defined as an ordered sequence of random numbers with respect to time. More formally, such a *stochastic process* can be written as

$$\{y(s,t), s \in \mathfrak{S}, t \in \mathfrak{T}\}, \tag{1.1}$$

where, for each $t \in \mathfrak{T}$, $y(\cdot,t)$ is a random variable on the sample space $\mathfrak{S}$, and a realization of this stochastic process is given by $y(s,\cdot)$ for each $s \in \mathfrak{S}$ with regard to a point in time $t \in \mathfrak{T}$. Hence, what we observe in reality are realizations of an unknown stochastic process, the *data-generating process*

$$\{y\}_{t=1}^{T} = \{y_1, y_2, \dots, y_t, \dots, y_{T-1}, y_T\} \tag{1.2}$$

with $t = 1, \dots, T \in \mathfrak{T}$.

One aim of time series analysis is concerned with the detection of this data-generating process by inferring from its realization to the underlying structure. In Figure 1.1, the path of real U.S. gross national product in billions of dollars (GNP) is depicted.[2] By mere eye-spotting, a "trend" in the series is evident. By comparing the behavior of this series with the unemployment rate for the same time span (*i.e.*, from 1909 until 1988), a lack of a "trend" is visible.

This artifact leads us to the first characteristic of a time series, namely *stationarity*. The ameliorated form of a stationary process is termed *weakly stationary* and is defined as

$$E[y_t] = \mu < \infty, \forall t \in \mathfrak{T}, \tag{1.3a}$$

$$E[(y_t - \mu)(y_{t-j} - \mu)] = \gamma_j, \forall t, j \in \mathfrak{T}. \tag{1.3b}$$

---

[1] The first occurrence of a subject entry is set in *italics*.

[2] The time series are taken from the extended Nelson and Plosser [1982] data set (see Schotman and van Dijk [1991]).

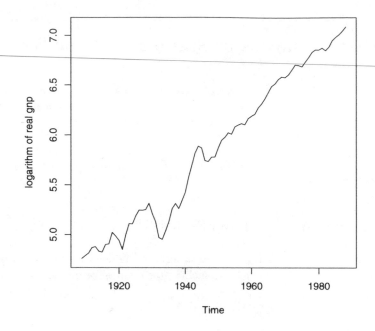

Fig. 1.1. U.S. GNP (real, logarithm)

Because only the first two theoretical moments of the stochastic process have to be defined and are constant and finite over time, this process is also referred to as being *second-order stationary* or *covariance stationary*. In that sense, the real U.S. GNP would not qualify as a realization of a stationary process because of its trending nature. Whether this is also the case for the U.S. unemployment rate (Figure 1.2) has to be seen.

Aside from weak stationarity, the concept of a *strictly stationary* process is defined as

$$F\{y_1, y_2, \ldots, y_t, \ldots, y_T\} = F\{y_{1+j}, y_{2+j}, \ldots, y_{t+j}, \ldots, y_{T+j}\}, \qquad (1.4)$$

where $F\{\cdot\}$ is the joint distribution function and $\forall t, j \in \mathfrak{T}$. Hence, if a process is strictly stationary with finite second moments, then it must be covariance stationary as well. A stochastic process can be covariance stationary without being strictly stationary. This would be the case, for example, if the mean and autocovariances were functions not of time but of higher moments instead.

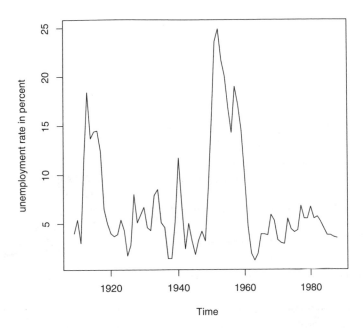

Fig. 1.2. U.S. unemployment rate (in percent)

The next characteristic of a stochastic process to be introduced is *ergodicity*.[3] Ergodicity refers to one type of asymptotic independence. In prose, asymptotic independence means that two realizations of a time series become ever closer to independence the further they are apart with respect to time. More formally, asymptotic independence can be defined as

$$|F(y_1, \ldots, y_T, y_{j+1}, \ldots, y_{j+T}) - F(y_1, \ldots, y_T)F(y_{j+1}, \ldots, y_{j+T})| \to 0 \quad (1.5)$$

with $j \to \infty$. The joint distribution of two sub-sequences of a stochastic process $\{y_t\}$ is closer to being equal to the product of the marginal distribution functions the more distant the two sub-sequences are from each other. A stationary stochastic process is ergodic if

$$\lim_{T \to \infty} \left\{ \frac{1}{T} \sum_{j=1}^{T} E[y_t - \mu][y_{t+j} - \mu] \right\} = 0 \quad (1.6)$$

---

[3] For a more detailed discussion and definition of ergodicity, the reader is referred to Davidson and MacKinnon [1993], Spanos [1986], White [1984], and Hamilton [1994].

holds. This equation would be satisfied if the autocovariances tended to zero with increasing j.

Finally, a *white noise* process is defined as

$$E(\varepsilon_t) = 0, \tag{1.7a}$$

$$E(\varepsilon_t^2) = \sigma^2, \tag{1.7b}$$

$$E(\varepsilon_t \varepsilon_\tau) = 0 \quad \text{for} \quad t \neq \tau. \tag{1.7c}$$

When necessary, ε_t is assumed to be normally distributed: $\varepsilon_t \backsim \mathcal{N}(0, \sigma^2)$. If Equations (1.7a)–(1.7c) are amended by this assumption, then the process is said to be a *normal* or *Gaussian white noise* process. Furthermore, sometimes Equation (1.7c) is replaced with the stronger assumption of independence. If this is the case, then the process is said to be an *independent white noise* process. Please note that for normally distributed random variables, uncorrelatedness and independence are equivalent. Otherwise, independence is sufficient for uncorrelatedness but not vice versa.

1.2 AR(p) Time Series Process

We start by considering a simple *first-order autoregressive process*. The current period's value of $\{y_t\}$ is explained by its previous one, a constant c, and an error process $\{\varepsilon_t\}$,

$$y_t = c + \phi y_{t-1} + \varepsilon_t, \tag{1.8}$$

where $\{\varepsilon_t\}$ obeys Equations (1.7a)–(1.7c); *i.e.*, it is a white noise process. Basically, Equation (1.8) is a first-order inhomogeneous difference equation. The path of this process depends on the value of ϕ. If $|\phi| \geq 1$, then shocks accumulate over time and hence the process is non-stationary. Incidentally, if $|\phi| > 1$, the process grows without bounds, and if $|\phi| = 1$ is true, the process has a *unit root*. The latter will be discussed in more detail in Section 3.2. In this section, however, we will only consider the covariance-stationary case, $|\phi| < 1$. With the lag operator L, Equation (1.8) can be rewritten as

$$(1 - \phi L)y_t = c + \varepsilon_t. \tag{1.9}$$

The stable solution to this process is given by an infinite sum of past errors with decaying weights:

$$y_t = (c + \varepsilon_t) + \phi(c + \varepsilon_{t-1}) + \phi^2(c + \varepsilon_{t-2}) + \phi^3(c + \varepsilon_{t-3}) + \ldots \tag{1.10a}$$

$$= \left[\frac{c}{1 - \phi} \right] + \varepsilon_t + \phi \varepsilon_{t-1} + \phi^2 \varepsilon_{t-2} + \phi^3 \varepsilon_{t-3} + \ldots . \tag{1.10b}$$

It is left to the reader as an exercise to show that the expected value and the second-order moments of the AR(1)-process in Equation (1.8) are given by

$$\mu = E[y_t] = \frac{c}{1 - \phi}, \tag{1.11a}$$

$$\gamma_0 = E[(y_t - \mu)^2] = \frac{\sigma^2}{1 - \phi^2}, \tag{1.11b}$$

$$\gamma_j = E[(y_t - \mu)(y_{t-j} - \mu)] = \left[\frac{\phi^j}{1 - \phi^2}\right]\sigma^2 \tag{1.11c}$$

(see Exercise 1). By comparing Equations (1.3a)–(1.3b) with (1.11a)–(1.11c), it is clear that the AR(1)-process $\{y_t\}$ is a stationary process. Furthermore, from Equation (1.11c), the geometrically decaying pattern of the autocovariances is evident.

In R code 1.1, a stable AR(1)-process with 100 observations and $\phi = 0.9$ is generated as well as a time series plot and its *autocorrelations* and *partial autocorrelations* as bar plots.[4] In Figure 1.3, the smooth behavior of $\{y_t\}$ caused by a value of ϕ close to one is visible. Also, the slowly decaying pattern of the autocorrelations is clearly given. The single spike at lag one in the partial autocorrelations indicates an AR(1)-process.

R Code 1.1 Simulation of AR(1)-process with $\phi = 0.9$

```
set.seed(123456)                                              1
y <- arima.sim(n = 100, list(ar = 0.9), innov=rnorm(100))    2
op <- par(no.readonly=TRUE)                                   3
layout(matrix(c(1, 1, 2, 3), 2, 2, byrow=TRUE))             4
plot.ts(y, ylab='')                                          5
acf(y, main='Autocorrelations', ylab='',                     6
    ylim=c(-1, 1), ci.col = "black")                         7
pacf(y, main='Partial Autocorrelations', ylab='',            8
    ylim=c(-1, 1), ci.col = "black")                         9
par(op)                                                     10
```

The AR(1)-process can be generalized to an AR(p)-process:

$$y_t = c + \phi_1 y_{t-1} + \phi_2 y_{t-2} + \ldots + \phi_p y_{t-p} + \varepsilon_t. \tag{1.12}$$

---

[4] In this R code example, functions contained in the standard package **stats** are used. However, it should be pointed out that the same functionalities are provided in the contributed CRAN package **fArma** by Würtz [2007a]. These functions include the simulation (`armaSim()`), estimation (`armaFit()`), and prediction (`predict()`) of autoregressive integrated moving average (ARIMA) models as well as stability evaluation (`armaRoots()`) and the calculation of theoretical autocorrelation and partial autocorrelation functions (`armaTrueacf()`).

Furthermore, S3 methods for summaries, printing, and plotting accompany these functions. The advantage for the user using these functions is given by a coherent argument list across all functions.

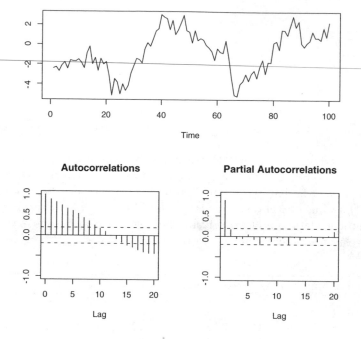

Fig. 1.3. Time series plot of AR(1)-process, $\phi = 0.9$

As with Equation (1.8), Equation (1.12) can be rewritten as

$$(1 - \phi_1 L - \phi_2 L^2 - \ldots - \phi_p L^p)y_t = c + \varepsilon_t. \tag{1.13}$$

It can be shown that such an AR(p)-process is stationary if all roots z_0 of the polynomial

$$\phi_p(z) = 1 - \phi_1 z - \phi_2 z^2 - \ldots - \phi_p z^p \tag{1.14}$$

have a modulus greater than one. The modulus of a complex number $z = z_1 + iz_2$ is defined as $|z| = \sqrt{z_1^2 + z_2^2}$. Viewing the stationarity condition from that point, it turns out that in the case of an AR(1)-process, as in Equation (1.8), $|\phi| < 1$ is required because the only solution to $1 - \phi z = 0$ is given for $z = 1/\phi$ and $|z| = |1/\phi| > 1$ when $|\phi| < 1$.

If the error process $\{\varepsilon_t\}$ is normally distributed, Equation (1.12) can be consistently estimated by the *ordinary least-squares* (OLS) method. Furthermore, the OLS estimator for the unknown coefficient vector $\boldsymbol{\beta} = (c, \boldsymbol{\phi})'$ is asymptotically normal. Alternatively, the model parameters can be estimated by the principle of maximum likelihood. However, one problem arises in the context of AR(p) models and this holds true for the more general class of ARMA(p, q) models discussed later. For independent and identically distributed random variables with probability density function $f(y_t; \boldsymbol{\theta})$ for $t = 1, \ldots, T$ and parameter vector $\boldsymbol{\theta}$, the joint density function is the product of the marginal densities,

$$f(\boldsymbol{y}; \boldsymbol{\theta}) = f(y_1, \ldots, y_T; \boldsymbol{\theta}) = \prod_{t=1}^{T} f(y_t; \boldsymbol{\theta}). \tag{1.15}$$

This joint density function can, in line with the maximum-likelihood principle, be interpreted as a function of the parameters $\boldsymbol{\theta}$ given the data vector $\boldsymbol{y}$; *i.e.*, the likelihood function is given as

$$\mathfrak{L}(\boldsymbol{\theta}|\boldsymbol{y}) = \mathfrak{L}(\boldsymbol{\theta}|y_1, \ldots, y_T) = \prod_{t=1}^{T} f(y_t; \boldsymbol{\theta}). \tag{1.16}$$

The log-likelihood function then has the simple form

$$\ln \mathfrak{L}(\boldsymbol{\theta}|\boldsymbol{y}) = \sum_{t=1}^{T} \ln f(y_t; \boldsymbol{\theta}). \tag{1.17}$$

Because our model assumes that the time series $\{y_t\}$ has been generated from a covariance-stationary process, the i.i.d. assumption is violated and hence the log-likelihood cannot be derived as swiftly as in Equations (1.15)–(1.17). That is, y_t is modeled as a function of its own history, and therefore y_t is not independent of $y_{t-1}, \ldots, y_{t-p}$ given that $\{\varepsilon_t\}$ is normally distributed with expectation $\mu = 0$ and variance σ^2. In order to apply the ML principle, one therefore has two options: either, estimate the full-information likelihood function or derive the likelihood function from a conditional marginal factorization. The derivation of the log-likelihood for both options is provided, for instance, in Hamilton [1994]. Here, we will focus on the second option. The idea is that the joint density function can be factored as the product of the conditional density function given all past information and the joint density function of the initial values,

$$f(y_T, \ldots, y_1; \boldsymbol{\theta}) = \left(\prod_{t=p+1}^{T} f(y_t|\mathcal{I}_{t-1}, \boldsymbol{\theta}) \right) \cdot f(y_p, \ldots, y_1; \boldsymbol{\theta}), \tag{1.18}$$

where $\mathcal{I}_{t-1}$ signifies the information available at time t. This joint density function can then be interpreted as the likelihood function with respect to the parameter vector $\boldsymbol{\theta}$ given the sample $\boldsymbol{y}$, and therefore the log-likelihood is given as

$$\ln \mathfrak{L}(\boldsymbol{\theta}|\boldsymbol{y}) = \sum_{t=p+1}^{T} \ln f(y_t|\mathcal{I}_{t-1}, \boldsymbol{\theta}) + \ln f(y_p, \ldots, y_1; \boldsymbol{\theta}). \tag{1.19}$$

The log-likelihood consists of two terms. The first term signifies the conditional log-likelihood and the second term the marginal log-likelihood for the initial values. Whether one maximizes the exact log-likelihood as in Equation (1.19) or only the conditional log-likelihood (*i.e.*, the first term of the exact log-likelihood) is asymptotically equivalent. Both are consistent estimators and have the same limiting normal distribution. Please bear in mind that in small samples the two estimators might differ by a non-negligible amount, in particular if the roots are close to unity. A derivation of the exact and conditional log-likelihood functions can be found for instance in Hamilton [1994]. Because a closed-form solution does not exist, numerical optimization methods are employed for deriving optimal parameter values.

In Figure 1.4, an AR(2)-process is displayed and generated according to

$$y_t = 0.6y_{t-1} - 0.28y_{t-2} + \varepsilon_t. \tag{1.20}$$

The *stability* of such a process can easily be checked with the function poly-root(). In R code 1.2, this AR(2)-process is generated by using the function filter() instead of arima.sim() as in R code 1.1 (see command line 2). The advantage of using filter() is that unstable AR(p)-processes can also be generated. Next, the AR(2)-process generated is estimated with the function arima(). The estimates are close to their theoretical values, as could be expected with a sample size of 1000. The moduli of the characteristic polynomial are retrieved with Mod() and the real and complex parts with the functions Re() and Im(), respectively. Please note that the signs of the estimated coefficients have to be reversed for the calculation of the roots (see command lines 7 and 8). The roots can be depicted in a Cartesian coordinate system with a *unit circle*, as is shown in Figure 1.5 on page 13.

1.3 MA(q) Time Series Process

It was shown in Section 1.2 that a finite stable AR(p)-process can be inverted to a moving average of contemporaneous and past shocks (see Equations (1.10a) and (1.10b)). We consider now how a process can be modeled as a finite *moving average* of its shocks. Such a process is called MA(q), where the parameter q refers to the highest lag of shocks to be included in such a process. We do so by first analyzing an MA(1)-process,

$$y_t = \mu + \varepsilon_t + \theta\varepsilon_{t-1}, \tag{1.21}$$

where $\{\varepsilon_t\}$ is a white noise process and μ, θ can be any constants. The moments of this MA(1)-process are given by

$$\mu = E[y_t] = E[\mu + \varepsilon_t + \theta\varepsilon_{t-1}], \tag{1.22a}$$

$$\gamma_0 = E[(y_t - \mu)^2] = (1 + \theta^2)\sigma^2, \tag{1.22b}$$

$$\gamma_1 = E[(y_t - \mu)(y_{t-1} - \mu)] = \theta\sigma^2. \tag{1.22c}$$

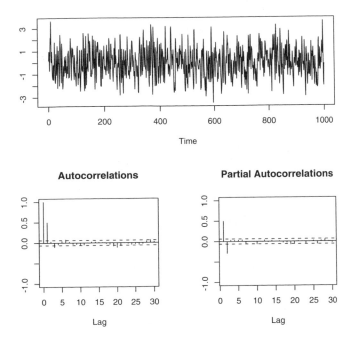

Fig. 1.4. Time series plot of AR(2)-process, $\phi_1 = 0.6$ and $\phi_2 = -0.28$

It is left to the reader to show that the higher autocovariances γ_j with $j > 1$ are nil. Neither the mean nor the autocovariance are functions of time, and hence an MA(1)-process is covariance stationary for all values of θ. Incidentally, because Equation (1.6) is satisfied, this process also has the characteristic of ergodicity.

Similar to R code 1.1, an MA(1)-process has been generated with $\mu = 0$ and $\theta = 0.8$ and is displayed in Figure 1.6. Let us now extend the MA(1)-process to the general class of MA(q)-processes

$$y_t = \mu + \varepsilon_t + \theta_1 \varepsilon_{t-1} + \ldots + \theta_q \varepsilon_{t-q}. \tag{1.23}$$

With the lag operator L, this process can be rewritten as

$$y_t - \mu = \varepsilon_t + \theta_1 \varepsilon_{t-1} + \ldots \theta_q \varepsilon_{t-q} \tag{1.24a}$$

$$= (1 + \theta_1 L + \ldots + \theta_q L^q)\varepsilon_t = \theta_q(L)\varepsilon_t. \tag{1.24b}$$

Similar to the case in which a stable AR(p)-process can be rewritten as an infinite MA-process, an MA(q)-process can be transformed to an infinite AR-process as long as the roots of the characteristic polynomial, the z-transform, have modulus greater than one (*i.e.*, are outside the unit circle):

R Code 1.2 Estimation of AR(2)-process with $\phi_1 = 0.6$ and $\phi_2 = -0.28$

```
series <-  rnorm(1000)                                              1
y.st <- filter(series, filter=c(0.6, -0.28),                       2
           method='recursive')                                      3
ar2.st <- arima(y.st, c(2, 0, 0), include.mean=FALSE,              4
           transform.pars=FALSE, method="ML")                       5
ar2.st$coef                                                         6
polyroot(c(1, -ar2.st$coef))                                        7
Mod(polyroot(c(1, -ar2.st$coef)))                                   8
root.comp <- Im(polyroot(c(1, -ar2.st$coef)))                      9
root.real <- Re(polyroot(c(1, -ar2.st$coef)))                      10
# Plotting the roots in a unit circle                              11
x <- seq(-1, 1, length = 1000)                                     12
y1 <- sqrt(1- x^2)                                                 13
y2 <- -sqrt(1- x^2)                                                14
plot(c(x, x), c(y1, y2), xlab='Real part',                        15
      ylab='Complex part', type='l',                              16
      main='Unit Circle', ylim=c(-2, 2), xlim=c(-2, 2))           17
abline(h=0)                                                        18
abline(v=0)                                                        19
points(Re(polyroot(c(1, -ar2.st$coef))),                          20
         Im(polyroot(c(1, -ar2.st$coef))), pch=19)                21
legend(-1.5, -1.5, legend="Roots of AR(2)", pch=19)               22
```

$$\theta_q z = 1 + \theta_1 z + \ldots + \theta_q z^q. \tag{1.25}$$

The expected value of an MA(q)-process is μ and hence invariant with respect to its order. The second-order moments are given as

$$\gamma_0 = E[(y_t - \mu)^2] = (1 + \theta_1^2 + \ldots + \theta_q^2)\sigma^2, \tag{1.26a}$$

$$\gamma_j = E[(\varepsilon_t + \theta_1\varepsilon_{t-1} + \ldots + \theta_q\varepsilon_{t-q})$$
$$\times (\varepsilon_{t-q} + \theta_1\varepsilon_{t-j-1} + \ldots + \theta_q\varepsilon_{t-j-q})]. \tag{1.26b}$$

Because $\{\varepsilon_t\}$ are uncorrelated with each other by assumption, Equation (1.26b) can be simplified to

$$\gamma_j = \begin{cases} (1 + \theta_{j+1}\theta_1 + \theta_{j+2}\theta_2 + \ldots + \theta_q\theta_{q-j})\sigma^2 & \text{for } j = 1, 2, \ldots, q \\ 0 & \text{for } j > q. \end{cases} \tag{1.27}$$

That is, empirically an MA(q)-process can be detected by its first q significant *autocorrelations* and a slowly decaying or alternating pattern of its *partial autocorrelations*. For large sample sizes T, a 95% significance band can be calculated as

$$\left(\varrho_j - \frac{2}{\sqrt{T}}, \varrho_j + \frac{2}{\sqrt{T}}\right), \tag{1.28}$$

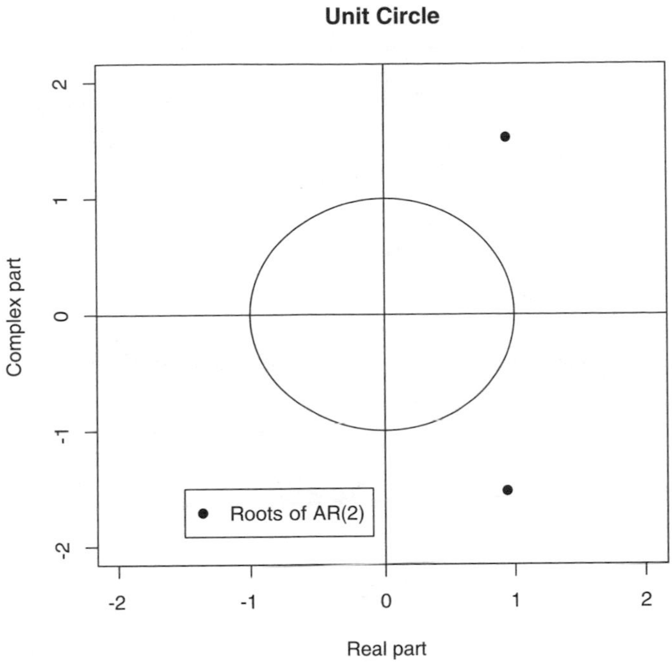

Unit Circle

Fig. 1.5. Unit circle and roots of stable AR(2)-process, $\phi_1 = 0.6$ and $\phi_2 = -0.28$

where ϱ_j refers to the jth-order autocorrelation.

It has been shown in Equation (1.10) that a finite AR-process can be inverted to an infinite MA-process. Before we proceed further, let us first examine the *stability* condition of such an MA(∞)-process,

$$y_t = \mu + \sum_{j=0}^{\infty} \psi_j \varepsilon_{t-j}. \tag{1.29}$$

Now, we ascribe the coefficients for an infinite process as ψ instead of θ, which was the case for MA(q)-processes. It can be shown that such an infinite process is covariance stationary if the coefficient sequence $\{\psi_j\}$ is either square summable,

$$\sum_{j=0}^{\infty} \psi_j^2 < \infty, \tag{1.30}$$

or absolute summable,

$$\sum_{J=0}^{\infty} |\psi_j| < \infty, \tag{1.31}$$

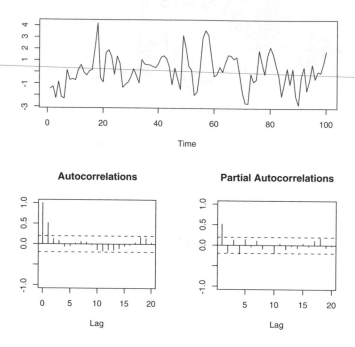

Fig. 1.6. Time series plot of MA(1)-process, $\theta = 0.8$

where absolute summability is sufficient for square summability; *i.e.*, the former implies the latter, but not vice versa.

1.4 ARMA(p, q) Time Series Process

It has been shown in the last two sections how a time series can be explained either by its history or by contemporaneous and past shocks. Furthermore, the moments of these data-generating processes have been derived and the mutual invertibility of these model classes has been stated for parameter sets that fulfill the stability condition. In this section, these two time series processes are put together; hence, a more general class of ARMA(p, q)-processes is investigated.

In practice, it is often cumbersome to detect a pure AR(p)- or MA(q)-process by the behavior of its empirical autocorrelation and partial autocorrelation functions because neither one tapers off with increasing lag order. In these instances, the time series might have been generated by a mixed *autoregressive moving average process*.

For a *stationary* time series $\{y_t\}$, such a mixed process is defined as

$$y_t = c + \phi_1 y_{t-1} + \ldots + \phi_p y_{t-p} + \varepsilon_t + \theta_1 \varepsilon_{t-1} + \ldots \theta_q \varepsilon_{t-q}. \tag{1.32}$$

By assumption, $\{y_t\}$ is *stationary*; *i.e.*, the roots of the *characteristic polynomial* lie outside the unit circle. Hence, with the lag operator, Equation (1.32) can be transformed to

$$y_t = \frac{c}{1 - \phi_1 L - \ldots - \phi_p L^p} + \frac{1 + \theta_1 L + \ldots + \theta_q L^q}{1 - \phi_1 L - \ldots - \phi_p L^p} \varepsilon_t \tag{1.33a}$$

$$= \mu + \psi(L)\varepsilon_t. \tag{1.33b}$$

The stated condition of absolute summability for the lag coefficients $\{\psi_j\}$ in Equation (1.31) must hold. Put differently, the stationarity condition depends only on the AR parameters and not on the moving average ones.

We will now briefly touch on the *Box-Jenkins* approach for time series modeling (see Box and Jenkins [1976]). This approach consists of three stages: identification, estimation, and diagnostic checking. As a first step, the series is visually inspected for stationarity. If an investigator has doubts that this condition is met, he or she has to suitably transform the series before proceeding. As we will see in Chapter 3, such transformations could be the removal of a deterministic trend or taking first differences with respect to time. Furthermore, variance instability such as higher fluctuations as time proceeds can be coped with by using the logarithmic values of the series instead. By inspecting of the empirical autocorrelation and partial autocorrelation functions, a tentative ARMA(p, q)-model is specified. The next stage is the estimation of a preliminary model. The employment of the ML principle allows one to discriminate between different model specifications by calculating information criteria and/or applying likelihood-ratio tests. Hence, one has at hand a second set of tools to determine an appropriate lag order for ARMA(p, q)-models compared with the order decision that is derived from ACF and PACF. Specifically, the Akaike [1981], Schwarz [1978], Hannan and Quinn [1979], and Quinn [1980] information criteria are defined as

$$\text{AIC}(p, q) = \ln(\hat{\sigma}^2) + \frac{2(p + q)}{T}, \tag{1.34}$$

$$\text{BIC}(p, q) = \ln(\hat{\sigma}^2) + \frac{\ln T (p + q)}{T}, \tag{1.35}$$

$$\text{HQ}(p, q) = \ln(\hat{\sigma}^2) + \frac{\ln(\ln(T))(p + q)}{T}, \tag{1.36}$$

where $\hat{\sigma}^2$ signifies the estimated variance of an ARMA(p, q)-process. The lag order (p, q) that minimizes the information criteria is then selected. As an alternative, a likelihood-ratio test can be computed for an unrestricted and a restricted model. The test statistic is defined as

$$2[\mathfrak{L}(\hat{\theta}) - \mathfrak{L}(\tilde{\theta})] \sim \chi^2(m), \tag{1.37}$$

where $\mathfrak{L}(\hat{\theta})$ denotes the unrestricted estimate of the log-likelihood and $\mathfrak{L}(\tilde{\theta})$ the one for the restricted log-likelihood. This test statistic is distributed as χ^2

with m degrees of freedom, which corresponds to the number of restrictions. Next, one should check the model's stability as well as the significance of its parameters. If one of these tests fails, the econometrician has to start anew by specifying a more parsimonious model with respect to the ARMA order. Now, let us assume that this is not the case. In the last step, diagnostic checking, he or she should then examine the residuals for uncorrelatedness and normality and conduct tests for correctness of the model's order *i.e.*, over- and underfitting. Incidentally, by calculating *pseudo ex ante* forecasts, the model's suitability for prediction can be examined.

As an example, we will apply the Box-Jenkins approach to the unemployment rate of the United States (see Figure 1.2).[5] Because no trending behavior is visible, we first examine its autocorrelation functions (see command lines 8 and 9 of R code 1.3). The graphs are displayed in Figure 1.7. The autocorrelation function tapers off, whereas the partial autocorrelation function has two significant correlations. As a tentative order, an ARMA(2, 0)-model is specified (see command line 13 of R code 1.3). This model is estimated with the function `arima()` contained in the package **stats**. The values of the estimated AR coefficients are $\phi_1 = 0.9297$ and $\phi_2 = -0.2356$. Their estimated standard errors are 0.1079 and 0.1077. Both AR coefficients are significantly different from zero, and the estimated values satisfy the stability condition. In the next step, the model's residuals are retrieved and stored in the object `res20`. As in the unemployment series, the residuals can be inspected visually, as can their autocorrelation functions (ACF) and partial autocorrelation functions (PACF). Furthermore, the assumption of uncorrelatedness can be tested with the *Ljung-Box Portmanteau test* (see Ljung and Box [1978]). This test is implemented in the `Box.test()` function of the package **stats**. Except for the PACF, these tools are graphically returned by the function `tsdiag()` (see Figure 1.8). The null hypothesis of uncorrelatedness up to order 20 cannot be rejected, given a marginal significance level of 0.3452. The hypothesis of normally distributed errors can be tested with the *Jarque-Bera test* for normality (see Bera and Jarque [1980] and Bera and Jarque [1981]), `jarque.bera.test()`, contained in the contributed CRAN package **tseries** by Trapletti and Hornik [2004], or with the *Shapiro-Wilk test* (see Shapiro and Wilk [1965] and Shapiro, Wilk and Chen [1968]), `shapiro.test()`, for example. Given a p-value of 0.9501, the normality hypothesis cannot be rejected. It should be noted that the assumption of normality could be visually inspected by a normal *quantiles plot* (`qqnorm()`). The value of the log-likelihood is -48.59, and the AIC takes a value of 105.18. The former value can be obtained by applying the `logLik` method to objects with class attribute **Arima**, and the latter is a list element of the returned object. In the next step, an overparametrized ARMA(3, 0)-model is estimated. It turns out that first the coefficient for the third lag is not significantly different from zero and that

---

[5] We used the logarithmic values of the unemployment rate because of a changing variance with respect to time.

R Code 1.3 Box-Jenkins: U.S. unemployment rate

```
library(urca)                                                        1
data(npext)                                                          2
y <- ts(na.omit(npext$unemploy), start=1909, end=1988,               3
        frequency=1)                                                 4
op <- par(no.readonly=TRUE)                                          5
layout(matrix(c(1, 1, 2, 3), 2, 2, byrow=TRUE))                      6
plot(y, ylab="unemployment rate (logarithm)")                        7
acf(y, main='Autocorrelations', ylab='', ylim=c(-1, 1))             8
pacf(y, main='Partial Autocorrelations', ylab='',                    9
     ylim=c(-1, 1))                                                 10
par(op)                                                             11
## tentative ARMA(2,0)                                             12
arma20 <- arima(y, order=c(2, 0, 0))                               13
ll20 <- logLik(arma20)                                             14
aic20 <- arma20$aic                                                15
res20 <- residuals(arma20)                                         16
Box.test(res20, lag = 20, type =  "Ljung-Box")                    17
shapiro.test(res20)                                                18
## alternative specifications                                      19
## ARMA(3,0)                                                       20
arma30 <- arima(y, order=c(3, 0, 0))                               21
ll30 <- logLik(arma30)                                             22
aic30 <- arma30$aic                                                23
lrtest <- as.numeric(2*(ll30 - ll20))                             24
chi.pval <- pchisq(lrtest, df = 1, lower.tail = FALSE)            25
## ARMA(1,1)                                                       26
arma11 <- arima(y, order = c(1, 0, 1))                            27
ll11 <- logLik(arma11)                                            28
aic11 <- arma11$aic                                               29
tsdiag(arma11)                                                    30
res11 <- residuals(arma11)                                        31
Box.test(res11, lag = 20, type =  "Ljung-Box")                   32
shapiro.test(res11)                                               33
tsdiag(arma11)                                                    34
## Using auto.arima()                                             35
library(forecast)                                                 36
auto.arima(y, max.p = 3, max.q = 3, start.p = 1,                 37
           start.q = 1, ic = "aic")                               38
```

---

second the estimates for the first- and second-order AR-coefficients remain
almost unchanged. However, the value of the log-likelihood is -47.47, and the
AIC is 104.93. Both indicate that an ARMA(3, 0)-model should be favored
compared with the ARMA(2, 0) specification, but the improvement in the
log-likelihood is not significant given a p-value of the likelihood-ratio test of
0.134. Therefore, one would prefer the more parsimonious ARMA(2, 0)-model

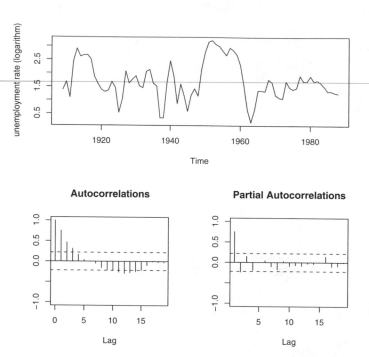

Fig. 1.7. Time series plot, ACF, and PACF of U.S. unemployment rate

over the ARMA(3, 0) specification. As a second alternative, an ARMA(1, 1)-model is specified. For this model, the values of the log-likelihood (-46.51) and the AIC (101.01) are more favorable compared with the ARMA(2, 0) specification. Furthermore, the diagnostic tests do not indicate any misspecification, and one would now prefer this model on the grounds of a higher value of the log-likelihood. Incidentally, the same specification is achieved by using the function `auto.arima()` contained in the package **forecast** (see Hyndman [2007]). Although the model's fit could be improved by including dummy variables to take into account the wide swings of the series during the pre–World War II era, by now we conclude that the U.S. unemployment rate can be well represented by an ARMA(1, 1)-model.

Once a stable (*i.e.*, covariance-stationary) ARMA(p, q)-model has been estimated, it can be used to predict future values of y_t. These forecasts can be computed recursively from the linear predictor

$$y_T(h) = \phi_1 \bar{y}_{T+h-1} + \ldots + \phi_p \bar{y}_{T+h-p} + \tag{1.38}$$

$$\varepsilon_t + \theta_1 \varepsilon_{t-T-1} + \ldots + \theta_q \varepsilon_{t-T-q} + \ldots, \tag{1.39}$$

where $\bar{y}_t = y_t$ for $t \leq T$ and $\bar{y}_{T+j} = y_T(j)$ for $j = 1, \ldots, h-1$. By employing the Wold representation of a covariance-stationary ARMA(p, q)-process (see

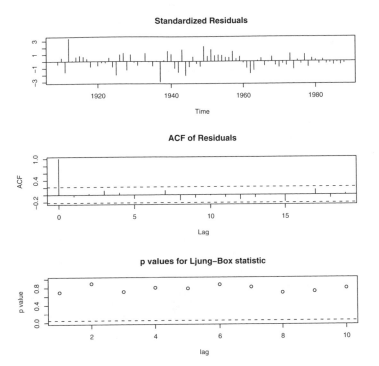

Fig. 1.8. Time series plot, ACF, and Ljung-Box values of residuals [ARMA(1, 1)]

Equations (1.33a) and (1.33b)), this predictor is equivalent to

$$y_T(h) = \mu + \psi_h \varepsilon_t + \psi_{h+1}\varepsilon_{t-1} + \psi_{h+2}\varepsilon_{t-2} + \cdots. \qquad (1.40)$$

It can be shown that this predictor is minimal with respect to the mean squared error criterion based on the information set $\mathcal{I}_t$ (see, for instance, Judge, Griffiths, Hill, Lütkepohl and Lee [1985, Chapter 7] and Hamilton [1994, Chapter 4]). Please note that when the forecast horizon h is greater than the moving average order q, the forecasts are determined solely by the autoregressive terms in Equation (1.38).

In R code 1.4, the estimated ARMA(1, 1)-model is used to obtain forecasts 10 years ahead for the logarithm of the U.S. unemployment rate. Here, the predict methods for objects with class attribute Arima have been utilized.[6] In Figure 1.9, the forecasts and an approximate 95% confidence band are depicted. If $\{\varepsilon_t\}$ is assumed to be standard normally distributed, then it follows that the h-steps-ahead forecast is distributed as

$$y_{t+h}|\mathcal{I}_t \sim N\left(y_{t+h|t}, \sigma^2(1 + \psi_1^2 + \ldots + \psi_{h-1}^2)\right), \qquad (1.41)$$

---

[6] Alternatively, one could have used the function forecast() and its associated plot method contained in the package **forecast**.

R **Code 1.4** Box-Jenkins: Predictions of the U.S. unemployment rate

---

```
## Forecasts                                                     1
arma11.pred <- predict(arma11, n.ahead = 10)                     2
predict <- ts(c(rep(NA, length(y) - 1), y[length(y)],            3
               arma11.pred$pred), start = 1909,                  4
               frequency = 1)                                    5
upper <- ts(c(rep(NA, length(y) - 1), y[length(y)],              6
             arma11.pred$pred + 2 * arma11.pred$se),             7
             start = 1909, frequency = 1)                        8
lower <- ts(c(rep(NA, length(y) - 1), y[length(y)],              9
             arma11.pred$pred - 2 * arma11.pred$se),            10
             start = 1909, frequency = 1)                       11
observed <- ts(c(y, rep(NA, 10)), start=1909,                   12
               frequency = 1)                                   13
## Plot of actual and forecasted values                        14
plot(observed, type = "l",                                     15
     ylab = "Actual and predicted values", xlab = "")          16
lines(predict, col = "blue", lty = 2)                          17
lines(lower, col = "red", lty = 5)                             18
lines(upper, col = "red", lty = 5)                             19
abline(v = 1988, col = "gray", lty = 3)                        20
```

---

where ψ_i for $i = 1, \ldots, h - 1$ signifies the coefficients from the Wold representation of a covariance-stationary ARMA(p, q)-process. The 95% forecast confidence band can then be computed as

$$y_{t+h}|\mathcal{I}_t \pm 1.96 \cdot \sqrt{\sigma^2(1 + \psi_1^2 + \cdots + \psi_{h-1}^2)}. \qquad (1.42)$$

Summary

In this first chapter, the definition of a time series and the concept of its data-generating process have been introduced. You should now be familiar with how to characterize a time series by its moments and distinguish the different concepts of stationarity. Two model classes for a time series have been introduced, namely the autoregressive and the moving average models, as well as a combination thereof. You should be able to detect and distinguish the order of these models by investigating its autocorrelation and partial auto-correlation functions. Finally, the Box-Jenkins approach to time series analysis has been presented. It is decomposed into three stages: specification of a tentative model order, estimation, and diagnostic checking.

So far, we have restricted the presentation to stationary time series only. At first sight, this focus might seem to be too myopic given that many time series cannot be characterized by a stationary process, in particular in macroeconomic and financial data sets. Therefore, in Chapter 3, non-stationary time

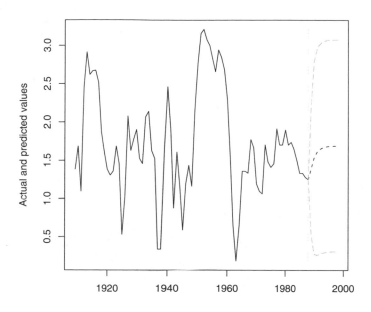

Fig. 1.9. Actual and forecasted values of the U.S. unemployment rate

series processes and how they can be transformed to achieve stationarity are discussed.

Exercises

1. Derive the second-order moments of an AR(1)-process as in Equation (1.8).
2. Generate stable AR(1)-, AR(2)-, and AR(3)-processes with $T = 1000$ for different error variances and plot their autocorrelations and partial autocorrelations. How could you determine the order of an AR(p)-process from its sample moments?
3. Show that the autocovariances $j > q$ of an MA(q)-process are zero.
4. Generate stable MA(1)-, MA(2)-, and MA(3)-processes with $T = 1000$ for different error variances and plot their autocorrelations and partial autocorrelations. How could you determine the order of an MA(q)-process from its sample moments?

2

Multivariate Analysis of Stationary Time Series

This is the second chapter that presents models confined to stationary time series, but now in the context of multivariate analysis. Vector autoregressive models and structural vector autoregressive models are introduced. The analytical tools of impulse response functions, forecast error variance decomposition, and Granger causality, as well as forecasting and diagnostic tests, are outlined. As will be shown later, these concepts can be applied to cointegrated systems, too.

2.1 Overview

Since the critique of Sims [1980] in the early 1980s, VAR analysis has evolved as a standard instrument in econometrics for analyzing multivariate time series. Because statistical tests are highly used in determining interdependence and dynamic relationships between variables, it soon became evident that this methodology could be enriched by incorporating non-statistical *a priori* information; hence SVAR models evolved that try to bypass these shortcomings. These kinds of models are considered in Section 2.3. At the same time as Sims jeopardized the paradigm of multiple structural equation models laid out by the Cowles Foundation in the 1940s and 1950s, Granger [1981] and Engle and Granger [1987] endowed econometricians with a powerful tool for modeling and testing economic relationships, namely the concept of integration and cointegration. Nowadays these traces of research are unified in the form of *vector error-correction* and *structural vector error-correction models*. These topics are deferred to Chapters 4 and 8.

2.2 Vector Autoregressive Models

2.2.1 Specification, Assumptions, and Estimation

In its basic form, a VAR consists of a set of K endogenous variables $\mathbf{y}_t = (y_{1t}, \ldots, y_{kt}, \ldots, y_{Kt})$ for $k = 1, \ldots K$. The VAR(p)-process is then defined as

$$\mathbf{y}_t = A_1 \mathbf{y}_{t-1} + \ldots + A_p \mathbf{y}_{t-p} + CD_t + \mathbf{u}_t, \tag{2.1}$$

where A_i are $(K \times K)$ coefficient matrices for $i = 1, \ldots, p$ and $\mathbf{u}_t$ is a K-dimensional white noise process with time-invariant positive definite covariance matrix $E(\mathbf{u}_t \mathbf{u}_t') = \Sigma_{\mathbf{u}}$. The matrix C is the coefficient matrix of potentially deterministic regressors with dimension $(K \times M)$, and D_t is an $(M \times 1)$ column vector holding the appropriate deterministic regressors, such as a constant, trend, and dummy and/or seasonal dummy variables.

Equation (2.1) is sometimes written in the form of a lag polynomial $A(L) = (I_K - A_1 - \ldots - A_p)$ as

$$A(L)\mathbf{y}_t = CD_t + \mathbf{u}_t. \tag{2.2}$$

One important characteristic of a VAR(p)-process is its stability. This means that it generates stationary time series with time-invariant means, variances, and covariance structure, given sufficient starting values. One can check this by evaluating the reverse characteristic polynomial,

$$\det(I_K - A_1 z - \ldots - A_p z^p) \neq 0 \quad \text{for } |z| \leq 1. \tag{2.3}$$

If the solution of the preceding equation has a root for $z = 1$, then either some or all variables in the VAR(p)-process are integrated of order one (*i.e.*, $I(1)$), a topic of the next chapter.

In practice, the stability of an empirical VAR(p)-process can be analyzed by considering the companion form and calculating the *eigenvalues* of the coefficient matrix (see Lütkepohl [2006] for a detailed derivation). A VAR(p)-process can be written as a VAR(1)-process as

$$\xi_t = A\xi_{t-1} + \mathbf{v}_t \tag{2.4}$$

with

$$\xi_t = \begin{bmatrix} \mathbf{y}_t \\ \vdots \\ \mathbf{y}_{t-p+1} \end{bmatrix}, \ A = \begin{bmatrix} A_1 & A_2 & \cdots & A_{p-1} & A_p \\ I & 0 & \cdots & 0 & 0 \\ 0 & I & \cdots & 0 & 0 \\ \vdots & \vdots & \ddots & \vdots & \vdots \\ 0 & 0 & \cdots & I & 0 \end{bmatrix}, \ \mathbf{v}_t = \begin{bmatrix} \mathbf{u}_t \\ 0 \\ \vdots \\ 0 \end{bmatrix}, \tag{2.5}$$

where the dimension of the stacked vectors ξ_t and $\mathbf{v}_t$ is $(Kp \times 1)$ and that of the matrix A is $(Kp \times Kp)$. If the moduli of the *eigenvalues* of A are less than one, then the VAR(p)-process is stable. For a given sample of the endogenous variables $\mathbf{y}_1, \ldots \mathbf{y}_T$ and sufficient presample values $\mathbf{y}_{-p+1}, \ldots, \mathbf{y}_0$, the coefficients of a VAR(p)-process can be estimated efficiently by least squares applied separately to each of the equations. If the error process $\mathbf{u}_t$ is normally distributed, then this estimator is equal to the maximum likelihood estimator conditional on the initial values.

It was shown in the previous chapter that a stable AR(p)-process can be represented as an infinite MA-process (see Equations (1.10a) and (1.10b)). This result applies likewise to a stable VAR(p)-process. Its *Wold moving average representation* is given as

$$\mathbf{y}_t = \Phi_0 \mathbf{u}_t + \Phi_1 \mathbf{u}_{t-1} + \Phi_2 \mathbf{u}_{t-2} + \ldots \tag{2.6}$$

with $\Phi_0 = I_K$, and the Φ_s matrices can be computed recursively according to

$$\Phi_s = \sum_{j=1}^{s} \Phi_{s-j} A_j \quad \text{for} \quad s = 1, 2, \ldots, \tag{2.7}$$

where $\Phi_0 = I_K$ and $A_j = 0$ for $j > p$.

Before considering an artificial data set, one topic should be touched on first, namely the empirical determination of an appropriate lag order. As in the univariate AR(p)-models, the lag length can be determined by *information criteria* such as those of Akaike [1981], Hannan and Quinn [1979], Quinn [1980], or Schwarz [1978], or by the *final prediction error* (see Lütkepohl [2006] for a detailed exposition of these criteria). These measures are defined as

$$\text{AIC}(p) = \log\det(\tilde{\Sigma}_u(p)) + \frac{2}{T} pK^2, \tag{2.8a}$$

$$\text{HQ}(p) = \log\det(\tilde{\Sigma}_u(p)) + \frac{2\log(\log(T))}{T} pK^2, \tag{2.8b}$$

$$\text{SC}(p) = \log\det(\tilde{\Sigma}_u(p)) + \frac{\log(T)}{T} pK^2, \text{ or} \tag{2.8c}$$

$$\text{FPE}(p) = \left(\frac{T+p^*}{T-p^*}\right)^K \det(\tilde{\Sigma}_u(p)), \tag{2.8d}$$

with $\tilde{\Sigma}_u(p) = T^{-1}\sum_{t=1}^{T} \hat{u}_t \hat{u}_t'$, and p^* is the total number of parameters in each equation and p assigns the lag order. It is shown in Lütkepohl [2006] that $\ln(\text{FPE})$ and AIC will indicate similar lag orders for moderate and large sample sizes. The following relations can be further deduced:

$$\hat{p}(\text{SC}) <= \hat{p}(\text{AIC}) \quad \text{if} \quad T >= 8, \tag{2.9a}$$

$$\hat{p}(\text{SC}) <= \hat{p}(\text{HQ}) \quad \text{for all} \quad T, \tag{2.9b}$$

$$\hat{p}(\text{HQ}) <= \hat{p}(\text{AIC}) \quad \text{if} \quad T >= 16. \tag{2.9c}$$

These information criteria are implemented in the functions VAR() and VARselect() contained in the package **vars**.[1] In the former function, an appropriate VAR(p)-model will be estimated by providing the maximal lag number, lag.max, and the desired criterion. The calculations are based upon the same sample size. That is, lag.max values are used as starting values for each of the estimated models. The result of the function VARselect() is a list object with elements **selection** and **criteria**. The element **selection** is a vector of optimal lag length according to the above-mentioned information criteria. The element **criteria** is a matrix containing the particular values for each of these criteria up to the maximal lag order chosen.

---

[1] The package **vars** can be obtained from CRAN, and it is hosted on R-Forge as project AICTS II; see http://CRAN.r-project.org and http://r-forge.r-project.org/projects/vars/, respectively.

Table 2.1. VAR result for y_1

Variable	Estimate	Std. Error	t-value	$\Pr(>|t|)$
Lagged levels				
y1.l1	0.4998	0.0354	14.1003	$0e+00$
y2.l1	0.1551	0.0407	3.8085	$2e-04$
y1.l2	-0.3291	0.0352	-9.3468	$0e+00$
y2.l2	-0.7550	0.0454	-16.6466	$0e+00$
Deterministic				
const.	5.9196	0.6197	9.5531	$0e+00$

We will now generate an artificial two-dimensional VAR(2)-process that obeys the following form:

$$\begin{bmatrix} y_1 \\ y_2 \end{bmatrix}_t = \begin{bmatrix} 5.0 \\ 10.0 \end{bmatrix} + \begin{bmatrix} 0.5 & 0.2 \\ -0.2 & -0.5 \end{bmatrix} \begin{bmatrix} y_1 \\ y_2 \end{bmatrix}_{t-1} + \begin{bmatrix} -0.3 & -0.7 \\ -0.1 & 0.3 \end{bmatrix} \begin{bmatrix} y_1 \\ y_2 \end{bmatrix}_{t-2} + \begin{bmatrix} u_1 \\ u_2 \end{bmatrix}_t . \quad (2.10)$$

The process above is simulated in R code 2.1. This is achieved by employing the function `ARMA()` and its method `simulate()`, contained in the package **dse1** (see Gilbert [2004], [2000], [1995], and [1993]).[2] In the first step, the lag polynomial $A(L)$ as described in Equation (2.2) is created as an array signified by `Apoly`. The shape of the variance-covariance matrix of the error process is an identity matrix stored as object `B`, and finally the constant term is assigned as `TRD`. An `ARMA` object is created next, and the model is simulated for a sample size of 500 observations. The resultant series are retrieved from the list element `output` and plotted in Figure 2.1. In the next step, the lag order is empirically determined by utilizing `VARselect()`. Alternatively, the VAR(p)-model could have been estimated directly by setting `lag.max = 4` and `type = "AIC"`. All criteria indicate a lag order of two. Finally, a VAR(2) with a constant is estimated with function `VAR()`, and its roots are checked for stability by applying the function `roots()` to the object `varsimest`. The function has an argument `"modulus"` of type logical that returns by default the moduli of the *eigenvalues*; otherwise a vector of complex numbers is returned.

The results of the VAR(2) for the variables y_1 and y_2 are presented in Tables 2.1 and 2.2, respectively. As expected, the estimated coefficients are close to their theoretical values, and all are significantly different from zero. Finally, the *eigenvalues* of the companion form are less than one and are provided in Table 2.3.

---

[2] Please note that this package is part of the bundle **dse**. As an alternative, a VAR-process can be simulated with the functions contained in the package **mAr**, too (see Barbosa [2007]).

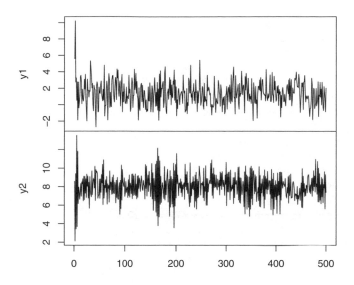

Fig. 2.1. Time series plot of the simulated VAR(2)-process

Table 2.2. VAR result for y_2

Variable	Estimate	Std. Error	t-value	$\Pr(>|t|)$
Lagged levels				
y1.l1	−0.1499	0.0358	−4.1920	$0e+00$
y2.l1	−0.4740	0.0411	−11.5360	$0e+00$
y1.l2	−0.1184	0.0355	−3.3328	$9e-04$
y2.l2	0.3006	0.0458	6.5684	$0e+00$
Deterministic				
const.	9.7620	0.6253	15.6124	$0e+00$

Table 2.3. Eigenvalues of the companion form

	1	2	3	4
Eigenvalues	0.8311	0.6121	0.6121	0.6049

R Code 2.1 Simulation of VAR(2)-process

```
## Simulate VAR(2)-data                                          1
library(dse1)                                                    2
library(vars)                                                    3
## Setting the lag-polynomial A(L)                               4
Apoly    <- array(c(1.0, -0.5, 0.3, 0,                           5
                    0.2, 0.1, 0, -0.2,                           6
                    0.7, 1, 0.5, -0.3) ,                         7
                  c(3, 2, 2))                                    8
## Setting Covariance to identity-matrix                         9
B <- diag(2)                                                     10
## Setting constant term to 5 and 10                             11
TRD <- c(5, 10)                                                  12
## Generating the VAR(2) model                                   13
var2   <- ARMA(A = Apoly, B = B, TREND = TRD)                    14
## Simulating 500 observations                                   15
varsim <- simulate(var2, sampleT = 500,                         16
                 noise = list(w = matrix(rnorm(1000),            17
nrow = 500, ncol = 2)), rng = list(seed = c(123456)))           18
## Obtaining the generated series                                19
vardat <- matrix(varsim$output, nrow = 500, ncol = 2)           20
colnames(vardat) <- c("y1", "y2")                               21
## Plotting the series                                           22
plot.ts(vardat, main = "", xlab = "")                           23
## Determining an appropriate lag-order                          24
infocrit <- VARselect(vardat, lag.max = 3,                      25
                    type = "const")                              26
## Estimating the model                                          27
varsimest <- VAR(vardat, p = 2, type = "const",                 28
                season = NULL, exogen = NULL)                    29
## Alternatively , selection according to AIC                    30
varsimest <- VAR(vardat, type = "const",                        31
                lag.max = 3, ic = "SC")                          32
## Checking the roots                                            33
roots <- roots(varsimest)                                        34
```

2.2.2 Diagnostic Tests

Once a VAR-model has been estimated, it is of pivotal interest to see whether the residuals obey the model's assumptions. That is, one should check for the absence of serial correlation and heteroscedasticity and see if the error process is normally distributed. In Section 1.4, these kinds of tests were briefly introduced, and the versions will now be presented in more detail for the multivariate case. As a final check, one can conduct structural stability tests; *i.e.*, CUSUM, CUSUM-of-squares, and/or fluctuation tests. The latter tests can be applied on a per-equation basis, whereas for the former tests multivariate statistics exist. All tests are made available in the package **vars**.

For testing the lack of serial correlation in the residuals of a VAR(p)-model, a Portmanteau test and the LM test proposed by Breusch [1978] and Godfrey [1978] are most commonly applied. For both tests, small sample modifications can be calculated, too, where the modification for the LM test was introduced by Edgerton and Shukur [1999]. The Portmanteau statistic is defined as

$$Q_h = T \sum_{j=1}^{h} \text{tr}(\hat{C}_j' \hat{C}_0^{-1} \hat{C}_j \hat{C}_0^{-1}) \tag{2.11}$$

with $\hat{C}_i = \frac{1}{T} \Sigma_{t=i+1}^{T} \hat{\mathbf{u}}_t \hat{\mathbf{u}}_{t-i}'$. The test statistic has an approximate $\chi^2(K^2 h - n^*)$ distribution, and n^* is the number of coefficients excluding deterministic terms of a VAR(p)-model. The limiting distribution is only valid for h tending to infinity at a suitable rate with growing sample size. Hence, the trade-off is between a decent approximation to the χ^2 distribution and a loss in power of the test when h is chosen too large. The small-sample properties of the test statistic

$$Q_h^* = T^2 \sum_{j=1}^{h} \frac{1}{T-j} \text{tr}(\hat{C}_j' \hat{C}_0^{-1} \hat{C}_j \hat{C}_0^{-1}) \tag{2.12}$$

may be better.

The Breusch-Godfrey LM-statistic is based upon the following auxiliary regressions:

$$\hat{\mathbf{u}}_t = A_1 \mathbf{y}_{t-1} + \ldots + A_p \mathbf{y}_{t-p} + CD_t + B_1 \hat{\mathbf{u}}_{t-1} + \ldots + B_h \hat{\mathbf{u}}_{t-h} + \varepsilon_t. \tag{2.13}$$

The null hypothesis is $H_0 : B_1 = \cdots = B_h = 0$, and correspondingly the alternative hypothesis is of the form $H_1 : \exists B_i \neq 0$ for $i = 1, 2, \ldots, h$. The test statistic is defined as

$$LM_h = T(K - \text{tr}(\tilde{\Sigma}_R^{-1} \tilde{\Sigma}_e)), \tag{2.14}$$

where $\tilde{\Sigma}_R$ and $\tilde{\Sigma}_e$ assign the residual covariance matrix of the restricted and unrestricted models, respectively. The test statistic LM_h is distributed as $\chi^2(hK^2)$. Edgerton and Shukur [1999] proposed a small-sample correction, which is defined as

$$LMF_h = \frac{1 - (1 - R_r^2)^{1/r}}{(1 - R_r^2)^{1/r}} \frac{Nr - q}{Km}, \tag{2.15}$$

with $R_r^2 = 1 - |\tilde{\Sigma}_e|/|\tilde{\Sigma}_R|$, $r = ((K^2 m^2 - 4)/(K^2 + m^2 - 5))^{1/2}$, $q = 1/2Km - 1$ and $N = T - K - m - 1/2(K - m + 1)$, where n is the number of regressors in the original system and $m = Kh$. The modified test statistic is distributed as $F(hK^2, int(Nr - q))$.

These tests are implemented in the function `serial.test()`. The test statistics are returned in the list element `serial` and have class attribute `htest`. Per default, the asymptotic Portmanteau test is returned. The adjusted version is computed if the `type` argument is set to `"PT.adjusted"`. The specifiers for the Breusch and Godfrey and the Edgerton and Shukur tests are `"BG"` and `"ES"`, respectively. The residuals are contained in the first list element. In R code 2.2, the asymptotic Portmanteau test is applied to the object `varsimest`.

R Code 2.2 Diagnostic tests of VAR(2)-process

```
## testing serial correlation                                    1
args(serial.test)                                                2
## Portmanteau-Test                                              3
var2c.serial <- serial.test(varsimest, lags.pt = 16,             4
                    type = "PT.asymptotic")                       5
var2c.serial                                                      6
plot(var2c.serial, names = "y1")                                 7
plot(var2c.serial, names = "y2")                                 8
## testing heteroscedasticity                                    9
args(arch.test)                                                 10
var2c.arch <- arch.test(varsimest, lags.multi = 5,             11
                    multivariate.only = TRUE)                   12
var2c.arch                                                     13
## testing for normality                                       14
args(normality.test)                                           15
var2c.norm <- normality.test(varsimest,                        16
                    multivariate.only = TRUE)                  17
var2c.norm                                                    18
## class and methods for diganostic tests                      19
class(var2c.serial)                                            20
class(var2c.arch)                                             21
class(var2c.norm)                                            22
methods(class = "varcheck")                                   23
## Plot of objects "varcheck"                                 24
args(vars:::plot.varcheck)                                    25
plot(var2c.serial, names = "y1")                             26
```

The implemented tests for heteroscedasticity are the univariate and multivariate ARCH tests (see Engle [1982], Hamilton [1994], and Lütkepohl [2006]). The multivariate ARCH-LM test is based on the following regression (the univariate test can be considered a special case of the exhibition below and is skipped):

$$vech(\hat{\mathbf{u}}_t\hat{\mathbf{u}}_t') = \beta_0 + B_1 vech(\hat{\mathbf{u}}_{t-1}\hat{\mathbf{u}}_{t-1}') + \ldots + B_q vech(\hat{\mathbf{u}}_{t-q}\hat{\mathbf{u}}_{t-q}') + \mathbf{v}_t, \quad (2.16)$$

where $\mathbf{v}_t$ assigns a spherical error process and *vech* is the column-stacking operator for symmetric matrices that stacks the columns from the main diagonal on downward. The *vech* operation is easily applied to a matrix by using `lower.tri(..., diag = TRUE)`. The dimension of β_0 is $\frac{1}{2}K(K+1)$, and for the coefficient matrices B_i with $i = 1, \ldots, q$, $\frac{1}{2}K(K+1) \times \frac{1}{2}K(K+1)$. The null hypothesis is $H_0 := B_1 = B_2 = \ldots = B_q = 0$ and the alternative is $H_1 : B_1 \neq 0 \cap B_2 \neq 0 \cap \ldots \cap B_q \neq 0$. The test statistic is defined as

$$\text{VARCH}_{LM}(q) = \frac{1}{2}TK(K+1)R_m^2, \tag{2.17}$$

with

$$R_m^2 = 1 - \frac{2}{K(K+1)}tr(\hat{\Omega}\hat{\Omega}_0^{-1}), \tag{2.18}$$

and $\hat{\Omega}$ assigns the covariance matrix of the regression model defined above. This test statistic is distributed as $\chi^2(qK^2(K+1)^2/4)$. These test statistics are implemented in the function `arch.test()` contained in the package **vars**. The default is to compute the multivariate test only. If `multivariate.only = FALSE`, the univariate tests are computed, too. In this case, the list object returned from `arch.test()` has three elements. The first element is the matrix of residuals. The second, signified by `arch.uni`, is a list object itself and holds the univariate test results for each of the series. The multivariate test result is contained in the third list element, signified by `arch.mul`. In R code 2.2, these tests are applied to the object `varsimest`.

The Jarque-Bera normality tests for univariate and multivariate series are implemented and applied to the residuals of a VAR(p) as well as separate tests for multivariate skewness and kurtosis (see Bera and Jarque [1980], [1981], Jarque and Bera [1987], and Lütkepohl [2006]). The univariate versions of the Jarque-Bera test are applied to the residuals of each equation. A multivariate version of this test can be computed by using the residuals that are standardized by a Choleski decomposition of the variance-covariance matrix for the centered residuals. Please note that in this case the test result is dependent upon the ordering of the variables. The test statistics for the multivariate case are defined as

$$JB_{mv} = s_3^2 + s_4^2, \tag{2.19}$$

where s_3^2 and s_4^2 are computed according to

$$s_3^2 = T\mathbf{b}_1'\mathbf{b}_1/6, \tag{2.20a}$$

$$s_4^2 = T(\mathbf{b}_2 - \mathbf{3}_K)'(\mathbf{b}_2 - \mathbf{3}_k)/24, \tag{2.20b}$$

and $\mathbf{b}_1$ and $\mathbf{b}_2$ are the third and fourth non-central moment vectors of the standardized residuals $\hat{\mathbf{u}}_t^s = \tilde{P}^-(\hat{\mathbf{u}}_t - \bar{\hat{\mathbf{u}}}_t)$ and $\tilde{P}$ is a lower triangular matrix with positive diagonal such that $\tilde{P}\tilde{P}' = \tilde{\Sigma}_{\mathbf{u}}$; *i.e.*, the Choleski decomposition of the residual covariance matrix. The test statistic JB_{mv} is distributed as $\chi^2(2K)$ and the multivariate skewness, s_3^2, and kurtosis test, s_4^2, are distributed as $\chi^2(K)$.

These tests are implemented in the function `normality.test()` contained in the package **vars**. Please note that the default is to compute the multivariate tests only. To obtain the test statistics for the single residual series, the argument `multivariate.only` has to be set to `FALSE`. The list elements of this function returned are `jb.uni` and `jb.mul`, which consist of objects with class attribute `htest` as for the previously introduced tests.

The three former functions return a list object with class attribute `varcheck` for which `plot` and `print` methods exist. The plots — one for each equation — include a residual plot, an empirical distribution plot, and the ACF and PACF of the residuals and their squares. The `plot` method offers additional arguments for adjusting its appearance. The residual plots as returned by `plot(var2c.norm)`, for instance, are provided in Figures 2.2 and 2.3 for y_1 and y_2, respectively. The results of the diagnostic tests are shown in Table 2.4.

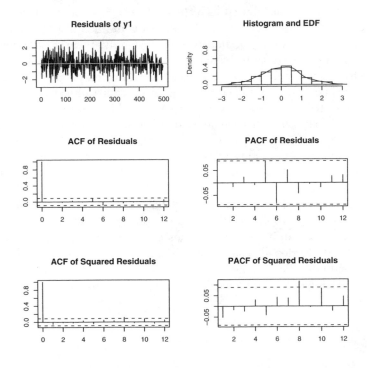

Fig. 2.2. Diagnostic residual plot for y_1 of VAR(2)-process

As expected for the simulated VAR(2)-process, none of the test outcomes indicate any deviations from a spherical error process.

Finally, structural stability can be tested by investigating the empirical fluctuation process. A detailed exposition of generalized fluctuation tests can be found for instance in Zeileis, Leisch, Hornik and Kleiber [2005] and Kuan

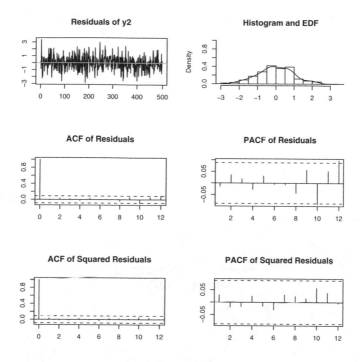

Fig. 2.3. Diagnostic residual plot for y_2 of VAR(2)-process

Table 2.4. Diagnostic tests of VAR(2)

Test	Statistic	D.F.	p-value
Portmanteau	52.44	56	0.61
ARCH VAR	32.58	45	0.92
JB VAR	0.54	4	0.97
Kurtosis	0.42	2	0.81
Skewness	0.12	2	0.94

and Hornik [1995]. Tests such as CUSUM, CUSUM-of-squares, MOSUM, and the fluctuation test are implemented in the function efp() contained in the package **strucchange**. The structural tests implemented in the package **strucchange** are explained in its vignette. The function stability() in the package **vars** is a wrapper function to efp(). The desired test is then applied to each of the equations in a VAR(p)-model. These kinds of tests are exhibited in R code 2.3, and their graphical results are depicted in Figures 2.4 and 2.5. In the code, an OLS-CUSUM and a fluctuation test have been applied to the simulated VAR(2)-process. In order to save space, only the test outcome for

y_1 (OLS-CUSUM test) and similarly the outcome for y_2 (fluctuation test) are shown. As expected, neither test indicates structural instability.

R Code 2.3 Empirical fluctuation processes

```
reccusum <- stability(varsimest,                    1
    type = "OLS-CUSUM")                             2
fluctuation <- stability(varsimest,                3
    type = "fluctuation")                          4
```

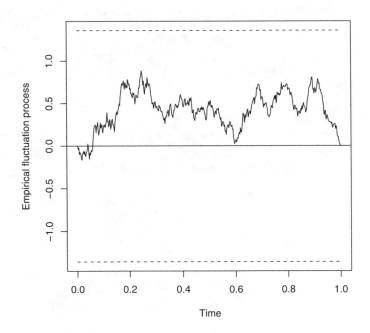

Fig. 2.4. OLS-CUSUM test for y_1 of VAR(2)-process

2.2.3 Causality Analysis

Often researchers are interested in the detection of causalities between variables. The most common one is the Granger causality test (see Granger [1969]). Incidentally, this test is not suited for testing causal relationships

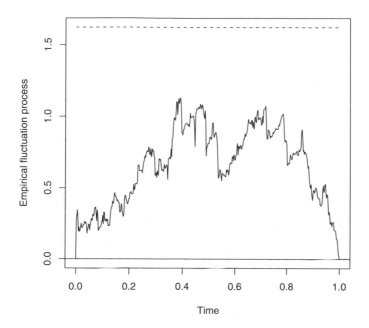

Fig. 2.5. Fluctuation test for y_2 of VAR(2)-process

in the strict sense because the possibility of a *post hoc ergo propter hoc* fallacy cannot be excluded. This is true for any "causality test" in econometrics. It is therefore common practice to say that variable x *granger-causes* variable y if variable x helps to predict variable y. Aside from this test, a Wald-type instantaneous causality test can be used, too. It is characterized by testing for non-zero correlation between the error processes of the cause and effect variables (see Lütkepohl [2006]).

For both tests, the vector of endogenous variables $\mathbf{y}_t$ is split into two sub-vectors $\mathbf{y}_{1t}$ and $\mathbf{y}_{2t}$ with dimensions $(K_1 \times 1)$ and $(K_2 \times 1)$ with $K = K_1 + K_2$. For the rewritten VAR(p),

$$\begin{bmatrix} \mathbf{y}_{1t} \\ \mathbf{y}_{2t} \end{bmatrix} = \sum_{i=1}^{p} \begin{bmatrix} \alpha_{11,i} \ \alpha_{12,i} \\ \alpha_{21,i} \ \alpha_{22,i} \end{bmatrix} \begin{bmatrix} \mathbf{y}_{1,t-i} \\ \mathbf{y}_{2,t-i} \end{bmatrix} + CD_t + \begin{bmatrix} \mathbf{u}_{1t} \\ \mathbf{u}_{2t} \end{bmatrix}, \qquad (2.21)$$

the null hypothesis that the sub-vector $\mathbf{y}_{1t}$ does not Granger-cause $\mathbf{y}_{2t}$ is defined as $\alpha_{21,i} = 0$ for $i = 1, 2, \ldots, p$. The alternative is $\exists \alpha_{21,i} \neq 0$ for $i = 1, 2, \ldots, p$. The test statistic is distributed as $F(pK_1K_2, KT - n^*)$, with n^* equal to the total number of parameters in the VAR(p)-process above, including deterministic regressors. The null hypothesis for non-instantaneous causality is defined as $H_0 : C\sigma = 0$, where C is an $(N \times K(K+1)/2)$ matrix

Table 2.5. Causality tests

Test	Statistic	p-value
Granger	250.07	0.00
Instant	0.00	0.99

of rank N selecting the relevant covariances of $\mathbf{u}_{1t}$ and $\mathbf{u}_{2t}$; $\tilde{\sigma} = vech(\tilde{\Sigma}_u)$. The Wald statistic is defined as

$$\lambda_W = T\tilde{\sigma}'C'[2CD_K^+(\tilde{\Sigma}_u \otimes \tilde{\Sigma}_u)D_K^{+'}C']^{-1}C\tilde{\sigma}, \tag{2.22}$$

where the Moore-Penrose inverse of the duplication matrix D_K is assigned by D_K^+ and $\tilde{\Sigma}_u = \frac{1}{T}\Sigma_{t=1}^T \hat{\mathbf{u}}_t\hat{\mathbf{u}}_t'$. The duplication matrix D_K has dimension $(K^2 \times \frac{1}{2}K(K+1))$ and is defined such that, for any symmetric $(K \times K)$ matrix A, $vec(A) = D_K vech(A)$ holds. The test statistic λ_W is asymptotically distributed as $\chi^2(N)$.

Both tests are implemented in the function `causality()` contained in the package **vars**. The function has two arguments. The first argument, x, is an object of class `varest`, and the second, cause, is a character vector of the variable names, which are assumed to be causal to the remaining variables in a VAR(p)-process. If this argument is unset, then the variable in the first column of x\$y is used as the cause variable and a warning is printed. In R code 2.4, this function is applied to the simulated VAR(2)-process. The results are provided in Table 2.5. Clearly, the null hypothesis of no Granger causality has to be dismissed, whereas the hypothesis of no instantaneous causality cannot be rejected.

R Code 2.4 Causality analysis of VAR(2)-process

```
## Causality tests                                              1
## Granger and instantaneous causality                         2
var.causal <- causality(varsimest, cause = "y2")               3
```

2.2.4 Forecasting

Once a VAR-model has been estimated and passes the diagnostic tests, it can be used for forecasting. Indeed, one of the primary purposes of VAR analysis is the detection of the dynamic interaction between the variables included in a VAR(p)-model. Aside from forecasts, other tools for investigating these relationships are impulse response analysis and forecast error variance decomposition, which will be covered in Subsections 2.2.5 and 2.2.6, respectively.

For a given empirical VAR, forecasts can be calculated recursively according to

$$\mathbf{y}_{T+h|T} = A_1\mathbf{y}_{T+h-1|T} + \ldots + A_p\mathbf{y}_{T+h-p|T} + CD_{T+h} \qquad (2.23)$$

for $h = 1, 2, \ldots, n$. The forecast error covariance matrix is given as

$$\mathrm{Cov}\left(\begin{bmatrix} \mathbf{y}_{T+1} - \mathbf{y}_{T+1|T} \\ \vdots \\ \mathbf{y}_{T+h} - \mathbf{y}_{T+h|T} \end{bmatrix}\right) = \begin{bmatrix} I & 0 & \cdots & 0 \\ \Phi_1 & I & & 0 \\ \vdots & & \ddots & 0 \\ \Phi_{h-1} & \Phi_{h-2} & \cdots & I \end{bmatrix} (\Sigma_{\mathbf{u}} \otimes I_h) \begin{bmatrix} I & 0 & \cdots & 0 \\ \Phi_1 & I & & 0 \\ \vdots & & \ddots & 0 \\ \Phi_{h-1} & \Phi_{h-2} & \cdots & I \end{bmatrix}',$$

and the matrices Φ_i are the coefficient matrices of the Wold moving average representation of a stable VAR(p)-process. Forecast confidence bands can then be calculated as

$$[y_{k,T+h|T} - c_{1-\gamma/2}\sigma_k(h), y_{k,T+h|T} + c_{1-\gamma/2}\sigma_k(h)], \qquad (2.24)$$

where $c_{1-\gamma/2}$ signifies the $(1 - \frac{\gamma}{2})$ percentage point of the normal distribution and $\sigma_k(h)$ is the standard deviation of the kth variable h steps ahead.

In the package **vars**, forecasting of VAR-processes is accomplished by a `predict` method for objects with class attribute **varest**. Besides the function's arguments for the **varest** object and the `n.ahead` forecast steps, a value for the forecast confidence interval can be provided, too. Its default value is 0.95. The `predict` method returns a list object of class **varprd** with three elements. The first element, `fcst`, is a list of matrices containing the predicted values, the lower and upper bounds according to the chosen confidence interval, `ci`, and its size. The second element, `endog`, is a matrix object containing the endogenous variables, and the third is the submitted **varest** object. A `plot` method for objects of class **varprd** exists as well as a `fanchart()` function for plotting *fan charts* as described in Britton, Fisher and Whitley [1998].

In R code 2.5, the `predict` method is applied to the empirical simulated VAR-process. The `fanchart()` function has `colors` and `cis` arguments, allowing the user to input vectors of colors and critical values. If these arguments are left `NULL`, then as defaults a gray color scheme is used and the critical values are set from 0.1 to 0.9 with a step size of 0.1. The predictions for y_1 are shown in Figure 2.6, and the fan chart for variable y_2 is depicted in Figure 2.7.

2.2.5 Impulse Response Functions

In Subsection 2.2.3, two causality tests were introduced, that are quite useful to infer whether a variable helps predict another one. However, this analysis

R Code 2.5 Forecasts of VAR-process

```
## Forecasting objects of class varest        1
args(vars ::: predict.varest)                 2
predictions <- predict(varsimest, n.ahead = 25,  3
                ci = 0.95)                     4
class(predictions)                            5
args(vars ::: plot.varprd)                    6
## Plot of predictions for y1                 7
plot(predictions, names = "y1")               8
## Fanchart for y2                            9
args(fanchart)                                10
fanchart(predictions, names = "y2")          11
```

Forecast of series y1

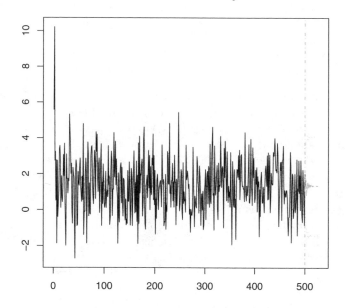

Fig. 2.6. Forecasting y_1 of VAR(2)-process

falls short of quantifying the impact of the impulse variable on the response variable over time. The *impulse response analysis* is used to investigate these kinds of dynamic interactions between the endogenous variables and is based upon the Wold moving average representation of a VAR(p)-process (see Equations (2.6) and (2.7)). The (i, j)th coefficients of the matrices Φ_s are thereby

Fanchart for variable y2

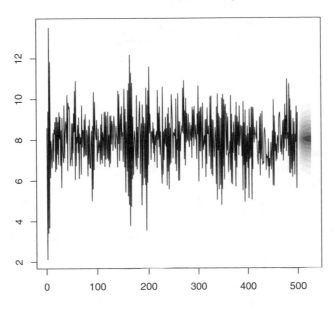

Fig. 2.7. Fanchart of y_2 of VAR(2)-process

interpreted as the expected response of variable $y_{i,t+s}$ to a unit change in variable y_{jt}. These effects can be cumulated through time $s = 1, 2, \ldots$, and hence one would obtain the cumulated impact of a unit change in variable j on the variable i at time s. Rather than these impulse response coefficients, it is often conceivable to use orthogonal impulse responses as an alternative. This is the case if the underlying shocks are less likely to occur in isolation but rather contemporaneous correlation between the components of the error process $\mathbf{u}_t$ exists; *i.e.*, the off-diagonal elements of $\Sigma_{\mathbf{u}}$ are non-zero. The orthogonal impulse responses are derived from a Choleski decomposition of the error variance-covariance matrix $\Sigma_{\mathbf{u}} = PP'$, with P being lower triangular. The moving average representation can then be transformed to

$$\mathbf{y}_t = \Psi_0 \varepsilon_t + \Psi_1 \varepsilon_{t-1} + \ldots, \qquad (2.25)$$

with $\varepsilon_t = P^{-1}\mathbf{u}_t$ and $\Psi_i = \Phi_i P$ for $i = 0, 1, 2, \ldots$ and $\Psi_0 = P$. Incidentally, because the matrix P is lower triangular, it follows that only a shock in the first variable of a VAR(p)-process exerts an influence on all the remaining ones and that the second and following variables cannot have a direct impact on y_{1t}. Hence, a certain structure of the error terms is implicitly imposed. One should bear this in mind when orthogonal impulse responses are employed.

Please note further that a different ordering of the variables might produce different outcomes with respect to the impulse responses. As we shall see in Section 2.3, the non-uniqueness of the impulse responses can be circumvented by analyzing a set of endogenous variables in the SVAR framework.

The function for conducting impulse response analysis is `irf()`, contained in the package **vars**. It is a method for objects with class attribute `varest`. The impulse variables are set as a character vector `impulse`, and the responses are provided likewise in the argument `response`. If either one is unset, then all variables are considered as impulses or responses, respectively. The default length of the impulse responses is set to 10 via argument `n.ahead`. The computation of orthogonal and/or cumulated impulse responses is controlled by the logical switches `ortho` and `cumulative`, respectively. Finally, confidence bands can be returned by setting `boot = TRUE` (default). The pre-set values are to run 100 replications and return 95% confidence bands. It is at the user's leisure to specify a `seed` for replicable results. The standard percentile interval is calculated as $CI_s = [s^*_{\gamma/2}, s^*_{(1-\gamma)/2}]$, where $s^*_{\gamma/2}$ and $s^*_{(1-\gamma)/2}$ are the $\gamma/2$ and $(1-\gamma)/2$ quantiles of the estimated bootstrapped impulse response coefficients $\hat{\Phi}^*$ or $\hat{\Psi}^*$ (see Efron and Tibshirani [1993]). The function `irf()` returns an object with class attribute `varirf` for which a `plot` and a `print` method exist.

In R code 2.6, an impulse response analysis is conducted for the simulated VAR(2)-process. For clarity, the impulse responses of y_1 to y_2 and vice versa have been split into two separate command lines. The results are shown in Figures 2.8 and 2.9, respectively.

R Code 2.6 IRA of VAR-process

```
## Impulse response analysis                            1
irf.y1 <- irf(varsimest, impulse = "y1",                2
              response = "y2", n.ahead = 10,            3
              ortho = FALSE, cumulative = FALSE,        4
              boot = FALSE, seed = 12345)               5
args(vars ::: plot.varirf)                              6
plot(irf.y1)                                            7
irf.y2 <- irf(varsimest, impulse = "y2",                8
              response = "y1", n.ahead = 10,            9
              ortho = TRUE, cumulative = TRUE,          10
              boot = FALSE, seed = 12345)               11
plot(irf.y2)                                            12
```

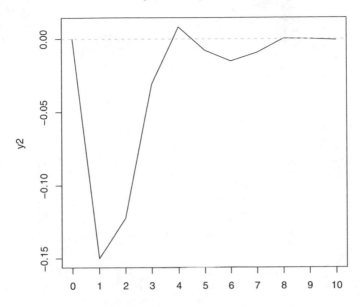

Fig. 2.8. Impulse responses of y_1 to y_2

2.2.6 Forecast Error Variance Decomposition

The forecast error variance decomposition (FEVD) is based upon the orthogonal impulse response coefficient matrices Ψ_n (see Subsection 2.2.4). The FEVD allows the user to analyze the contribution of variable j to the h-step forecast error variance of variable k. If the element-wise squared orthogonal impulse responses are divided by the variance of the forecast error variance, $\sigma_k^2(h)$, the result is a percentage figure. Formally, the forecast error variance for $y_{k,T+h} - Y_{k,T+h|T}$ is defined as

$$\sigma_k^2(h) = \sum_{n=0}^{h-1}(\psi_{k1,n}^2 + \ldots + \psi_{kK,n}^2), \tag{2.26}$$

which can be written as

$$\sigma_k^2(h) = \sum_{j=1}^{K}(\psi_{kj,0}^2 + \ldots + \psi_{kj,h-1}^2). \tag{2.27}$$

Dividing the term $(\psi_{kj,0}^2 + \ldots + \psi_{kj,h-1}^2)$ by $\sigma_k^2(h)$ yields the forecast error variance decompositions in percentage terms:

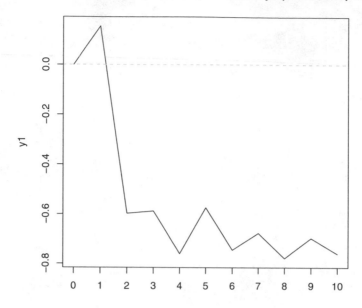

Fig. 2.9. Impulse responses of y_2 to y_1

$$\omega_{kj}(h) = (\psi_{kj,0}^2 + \ldots + \psi_{kj,h-1}^2)/\sigma_k^2(h). \tag{2.28}$$

The `fevd` method in the package **vars** is available for conducting FEVD. The argument `n.ahead` sets the number of forecasting steps; it has a default value of 10. In R code 2.7, an FEVD is applied to the simulated VAR(2)-process, and its graphical output is presented in Figure 2.10.

R Code 2.7 FEVD of VAR-process

```
## Forecast error variance decomposition          1
fevd.var2 <- fevd(varsimest, n.ahead = 10)        2
args(vars::: plot.varfevd)                         3
plot(fevd.var2, addbars = 2)                       4
```

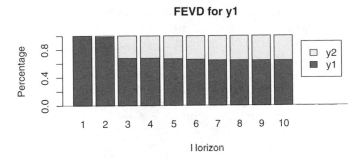

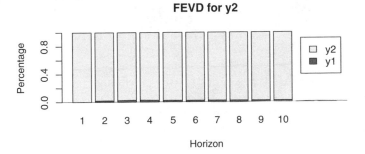

Fig. 2.10. FEVD for VAR(2)-process

2.3 Structural Vector Autoregressive Models

2.3.1 Specification and Assumptions

Recall from Subsection 2.2.1 the definition of a VAR(p)-process, in particular Equation (2.1). A VAR(p) can be interpreted as a reduced-form model. An SVAR model is its structural form and is defined as

$$Ay_t = A_1^* y_{t-1} + \ldots + A_p^* y_{t-p} + B\varepsilon_t. \qquad (2.29)$$

For a textbook exposition of SVAR-models, see Amisano and Giannini [1997]. It is assumed that the *structural errors*, ε_t, are white noise and the coefficient matrices A_i^* for $i = 1, \ldots, p$, are structural coefficients that will differ from their reduced-form counterparts if $A \neq I$. To see this, consider the resulting equation by left-multiplying Equation (2.29) with the inverse of A:

$$\begin{aligned} \mathbf{y}_t &= A^{-1}A_1^* \mathbf{y}_{t-1} + \ldots + A^{-1}A_p^* \mathbf{y}_{t-p} + A^{-1}B\varepsilon_t, \\ \mathbf{y}_t &= A_1 \mathbf{y}_{t-1} + \ldots + A_p \mathbf{y}_{t-p} + \mathbf{u}_t. \end{aligned} \qquad (2.30)$$

An SVAR-model can be used to identify shocks and trace these out by employing IRA and/or FEVD through imposing restrictions on the matrices A

and/or B. Incidentally, though an SVAR-model is a structural model, it departs from a reduced-form VAR(p)-model and only restrictions for A and B can be added. It should be noted that the reduced-form residuals can be retrieved from an SVAR-model by $\mathbf{u}_t = A^{-1}B\varepsilon_t$ and its variance-covariance matrix by $\Sigma_\mathbf{u} = A^{-1}BB'A^{-1'}$.

Depending on the restrictions imposed, three types of SVAR-models can be distinguished:

- A-model: B is set to I_K (minimum number of restrictions for identification is $K(K-1)/2$).
- B-model: A is set to I_K (minimum number of restrictions to be imposed for identification is the same as for A-model).
- AB-model: restrictions can be placed on both matrices (minimum number of restrictions for identification is $K^2 + K(K-1)/2$).

2.3.2 Estimation

Depending on the SVAR type, the estimation is similar to the estimation of a simultaneous multiple-equation model with covariance restrictions on the error terms. In practice, the maximum-likelihood principle is applied to the concentrated log-likelihood, which is given as

$$\ln L_c(A, B) = \text{const} + \frac{T}{2}\ln|A^2| - \frac{T}{2}\ln|B^2| - \frac{T}{2}\text{tr}(A'B'^{-1}B^{-1}A\hat{\Sigma}_u), \quad (2.31)$$

where $\hat{\Sigma}_u$ signifies the estimated residual covariance matrix of the VAR(p)-model. The negative of Equation (2.31) is minimized subject to the imposed restrictions on A and B, which can be compactly written as

$$\begin{bmatrix} vecA \\ vecB \end{bmatrix} = \begin{bmatrix} R_A & 0 \\ 0 & R_B \end{bmatrix} \begin{bmatrix} \gamma_A \\ \gamma_b \end{bmatrix} + \begin{bmatrix} r_A \\ r_B \end{bmatrix}. \quad (2.32)$$

Two approaches for numerically estimating the unknown coefficients are implemented within the R package **vars**. The first method applies the optim() function for direct minimization of the negative log-likelihood, whereas the second method makes use of the scoring algorithm proposed by Amisano and Giannini [1997]. Either method is selected by providing "direct" or "scoring" as the value for the argument estmethod in the function SVAR(). In addition, the first argument in a call to SVAR() must be an object of class varest. Whether an A-, B-, or AB-model will be estimated is dependent on the setting for Amat and Bmat. If a restriction matrix for Amat with dimension $(K \times K)$ is provided and the argument Bmat is left NULL, an A-model is estimated. In this case, Bmat is set to an identity matrix I_K. Alternatively, if only a matrix object for Bmat is provided and Amat is left unchanged, then a B-model will be estimated and internally Amat is set to an identity matrix I_K. Finally, if matrix objects for both arguments are provided, then an AB-model will be estimated. In all cases, the matrix elements to be estimated are marked by NA entries at the relevant positions. Depending on the chosen model, the list elements A, Ase, B, Bse contain the estimated coefficient matrices with the numerical standard errors, if applicable. In case estmethod = "direct", the standard errors are returned only if SVAR() has been called with hessian = TRUE. The returned list element Sigma.U is the variance-covariance matrix of the reduced-form residuals times 100; *i.e.*, $\Sigma_U = A^{-1}BB'A^{-1'} \times 100$. Please note that this estimated variance-covariance matrix only corresponds to the reduced-form counterpart if the SVAR-model is exactly identified. The validity of the overidentifying restrictions can be tested with an LR test defined as

$$LR = T(\ln|\tilde{\Sigma}_u| - \ln|\hat{\Sigma}_u|), \tag{2.33}$$

where $\tilde{\Sigma}_u$ is the implied variance-covariance matrix of the SVAR and $\hat{\Sigma}_u$ signifies the reduced-form counterpart. The statistic is distributed as χ^2 with degrees of freedom equal to the number of overidentifying restrictions. This test statistic is returned as list element LR with class attribute htest. The element opt is the object returned from function optim(). The remaining four list items are the vector of starting values, the SVAR-model type, the varest object, and the call to SVAR().

In R code 2.8, the function SVAR() is applied to a generated A-model of the form

$$\begin{bmatrix} 1 & -0.7 \\ 0.8 & 1 \end{bmatrix} \begin{bmatrix} y_1 \\ y_2 \end{bmatrix}_t = \begin{bmatrix} 0.5 & 0.2 \\ -0.2 & -0.5 \end{bmatrix} \begin{bmatrix} y_1 \\ y_2 \end{bmatrix}_{t-1} + \begin{bmatrix} -0.3 & -0.7 \\ -0.1 & 0.3 \end{bmatrix} \begin{bmatrix} y_1 \\ y_2 \end{bmatrix}_{t-2}$$
$$+ \begin{bmatrix} \varepsilon_1 \\ \varepsilon_2 \end{bmatrix}_t. \tag{2.34}$$

In the call to SVAR(), the argument hessian = TRUE has been used, which is passed to optim(). Hence, the empirical standard errors are returned in the list element Ase of the object svar.A. The result is shown in Table 2.6. The coefficients are close to their theoretical counterparts and statistically significant different from zero. As expected, the likelihood-ratio statistic for overidentification does not indicate the rejection of the null.

Table 2.6. SVAR A-model: estimated coefficients

Variable	y_1	y_2
y_1	1.0000	−0.6975
		(−13.67)
y_2	0.8571	1.0000
	(14.96)	

Note: t statistics in parentheses.

R Code 2.8 SVAR: A-model

```
library(dse1)                                                      1
library(vars)                                                      2
## A-model                                                         3
Apoly    <- array(c(1.0,  -0.5,  0.3,  0.8,                        4
                    0.2,  0.1,  -0.7,  -0.2,                       5
                    0.7,  1,  0.5,  -0.3) ,                        6
                  c(3, 2, 2))                                      7
## Setting covariance to identity-matrix                          8
B <- diag(2)                                                       9
## Generating the VAR(2) model                                    10
svarA   <- ARMA(A = Apoly, B = B)                                 11
## Simulating 500 observations                                    12
svarsim <- simulate(svarA, sampleT = 500,                        13
                    rng = list(seed = c(123456)))                14
## Obtaining the generated series                                15
svardat <- matrix(svarsim$output, nrow = 500, ncol = 2)         16
colnames(svardat) <- c("y1", "y2")                              17
## Estimating the VAR                                            18
varest <- VAR(svardat, p = 2, type = "none")                   19
## Setting up matrices for A-model                              20
Amat <- diag(2)                                                 21
Amat[2, 1] <- NA                                               22
Amat[1, 2] <- NA                                               23
## Estimating the SVAR A-type by direct maximisation           24
## of the log-likelihood                                       25
args(SVAR)                                                      26
svar.A <- SVAR(varest, estmethod = "direct",                  27
               Amat = Amat, hessian = TRUE)                    28
```

A B-type SVAR is first simulated and then estimated with the alternative method estmethod = "scoring" in R code 2.9. The scoring algorithm is based upon the updating equation

$$
\begin{bmatrix} \tilde{\gamma}_A \\ \tilde{\gamma}_B \end{bmatrix}_{i+1} = \begin{bmatrix} \tilde{\gamma}_A \\ \tilde{\gamma}_B \end{bmatrix}_i + \ell \mathcal{I} \left(\begin{bmatrix} \tilde{\gamma}_A \\ \tilde{\gamma}_B \end{bmatrix}_i \right)^{-1} \mathcal{S} \left(\begin{bmatrix} \tilde{\gamma}_A \\ \tilde{\gamma}_B \end{bmatrix}_i \right),
\tag{2.35}
$$

where ℓ signifies the step length, $\mathcal{I}$ is the information matrix for the unknown coefficients contained in $\tilde{\gamma}_A$ and $\tilde{\gamma}_B$, and $\mathcal{S}$ is the scoring vector. The iteration step is assigned by i.

The covariance for the error terms has been set to -0.8. The values of the coefficient matrices A_i for $i = 1, 2$ are the same as in the previous example. The result is provided in Table 2.7.

In addition to an object with class attribute varest, the other arguments of SVAR() if estmethod = "scoring" are max.iter for defining the maximal number of iterations, conv.crit for providing a value for defining convergence,

R Code 2.9 SVAR: B-model

```
library(dse1)                                                        1
library(vars)                                                       2
## B-model                                                         3
Apoly   <- array(c(1.0,  -0.5, 0.3, 0,                             4
                   0.2,  0.1, 0, -0.2,                             5
                   0.7,  1, 0.5, -0.3) ,                          6
                 c(3, 2, 2))                                       7
## Setting covariance to identity-matrix                           8
B <- diag(2)                                                        9
B[2, 1] <- -0.8                                                     10
## Generating the VAR(2) model                                      11
svarB   <- ARMA(A = Apoly, B = B)                                 12
## Simulating 500 observations                                     13
svarsim <- simulate(svarB, sampleT = 500,                         14
                    rng = list(seed = c(123456)))                 15
svardat <- matrix(svarsim$output, nrow = 500, ncol = 2)          16
colnames(svardat) <- c("y1", "y2")                                17
varest <- VAR(svardat, p = 2, type = "none")                     18
## Estimating the SVAR B-type by scoring algorithm                 19
## Setting up the restriction matrix and vector                    20
## for B-model                                                     21
Bmat <- diag(2)                                                    22
Bmat[2, 1] <- NA                                                   23
svar.B <- SVAR(varest, estmethod = "scoring",                    24
               Bmat = Bmat, max.iter = 200)                       25
```

Table 2.7. SVAR B-model: estimated coefficients

Variable	y_1	y_2
y_1	1.0000	0.0000
y_2	-0.8439	1.0000
	(-18.83)	

Note: t statistics in parentheses.

and maxls for determining the maximal step length. As in the estimation method direct, the alternative method returns an object with class attribute svarest. For objects of this class, methods for computing impulse responses and forecast error variance decomposition exist. These methods will be the subjects of the following two subsections.

2.3.3 Impulse Response Functions

Just as impulse response analysis can be conducted for objects with class attribute varest, it can also be done for objects with class attribute svarest (see Subsection 2.2.5 on page 37 following). In fact, the `irf` methods for classes `varest` and `svarest` are at hand with the same set of arguments, except `ortho` is missing for objects of class `svarest` due to the nature and interpretation of the error terms in an SVAR. The impulse response coefficients for an SVAR are calculated as $\Theta_i = \Phi_i A^{-1} B$ for $i = 1, \ldots, n$.

In R code 2.10, IRA is exhibited for the estimated A-type SVAR from the previous section. The impulses from `y1` to `y2` are calculated. In program line 3, the method is applied to the object `svar.A`. In line 6, these orthogonal impulse responses are plotted. The result is provided in Figure 2.11.

R Code 2.10 SVAR: Impulse response analysis

```
## Impulse response analysis of SVAR A-type model     1
args(vars ::: irf.svarest)                             2
irf.svara <- irf(svar.A, impulse = "y1",               3
                 response = "y2", boot = FALSE)         4
args(vars ::: plot.varirf)                             5
plot(irf.svara)                                         6
```

2.3.4 Forecast Error Variance Decomposition

A forecast error variance decomposition can be applied to objects of class `svarest`. Here the forecast errors of $y_{T+h|T}$ are derived from the impulse responses of an SVAR, and the derivation for the forecast error variance decomposition is similar to the one outlined for the VAR-model (see Subsection 2.2.6 on page 41 following).

R Code 2.11 SVAR: Forecast error variance decomposition

```
## FEVD analysis of SVAR B-type model      1
args(vars ::: fevd.svarest)                2
fevd.svarb <- fevd(svar.B, n.ahead = 5)    3
class(fevd.svarb)                          4
methods(class = "varfevd")                 5
plot(fevd.svarb)                           6
```

An application for the SVAR B-model is provided in R code 2.11. As for the FEVD for VAR-models, `print` and `plot` methods exist for SVAR-models. The outcome of the FEVD for the variable `y2` is provided in Table 2.8.

SVAR Impulse Response from y1

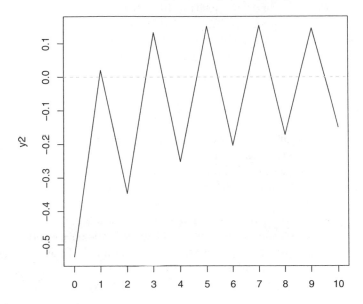

Fig. 2.11. IRA from y_1 to y_2 of SVAR A-model

Table 2.8. SVAR B-model: FEVD for y_2

Period	y_1	y_2
1	0.4160	0.5840
2	0.4021	0.5979
3	0.4385	0.5615
4	0.4342	0.5658
5	0.4350	0.5650

Summary

In this chapter, the analysis of stationary time series has been extended to multivariate models and their associated statistical tests and methods. In particular, VAR- and SVAR-models have been introduced, where the former can be interpreted as the reduced-form counterparts of SVAR-models. Both model classes have been illustrated by artificial data sets.

It has been outlined how a suitable lag length can be empirically determined and what kind of diagnostic tests are at hand for checking the assumptions about the multivariate error process. The different concepts of causality analysis and forecasting with VAR-models have been shown. For investigating the dynamic interactions between variables, the impulse response functions and forecast error variance decomposition have been introduced. These tools are implemented as methods for VAR- and SVAR-models alike. The results can be obtained and plotted swiftly with the functions included in package **vars**. An overview of the package's structure is presented in Table 2.9.

Exercises

1. Set up a three-dimensional VAR(2) model where the third variable does not Granger-cause the first variable.
2. Simulate 250 observations of your model from Exercise 1.
3. Estimate a VAR(2)-model with the simulated data from Exercise 2 and check its stability.
4. Conduct the diagnostic tests outlined in Subsection 2.2.2.
5. Perform Granger-causality tests for y_3, Granger-causing y_2 and y_1.
6. Calculate the impulse response functions (orthogonal and non-orthogonal) and forecast error variance decomposition for y_3.

Table 2.9. Overview of package **vars**

function or method class		methods for class	functions for class
VAR	varest	coef, fevd, fitted, irf, logLik, Phi, plot, predict, print, Psi, resid, summary	Acoef, arch.test, Bcoef, BQ, causality, normality.test, restrict, roots, serial.test, stability
SVAR	svarest	fevd, irf, logLik, Phi, print, summary	
SVEC	svecest	fevd, irf, logLik, Phi, print, summary	
vec2var	vec2var	fevd, fitted, irf, logLik, Phi, predict, print, Psi, resid	arch.test, normality.test, serial.test
fevd	varfevd	plot, print	
irf	varirf	plot, print	
predict	varprd	plot, print	fanchart
summary	varsum, svarsum, svecsum	print	
arch.test	varcheck	plot, print	
normality.test	varcheck	plot, print	
serial.test	varcheck	plot, print	
stability	varstabil	plot, print	

7. Set up an SVAR-model of type AB with three variables and two lags, which are just identified and overidentified, respectively.
8. Simulate 250 observations of your model from Exercise 7 and estimate it with function SVAR2.
9. Perform impulse response analysis and forecast error variance decomposition of your estimated SVAR AB-model.

3

Non-stationary Time Series

In this chapter, models for non-stationary time series are introduced. Before the characteristics of unit processes are presented, the differences between trend- and difference-stationary models are outlined. In the last section, long-memory processes (i.e., fractionally integrated processes) are presented as a bridge between stationary and unit root processes.

3.1 Trend- versus Difference-Stationary Series

In the first chapter, a model class for univariate, stationary time series was introduced. For instance, it has been shown that a stable autoregressive process ($AR(p)$) can be inverted to an infinite moving average process with a constant mean. However, most macroeconomic time series seem not to adhere to such a data-generating process (see Figure 1.1). In this section, we will consider a more encompassing data-generating process that was presented by Campbell and Perron [1991].

Now, it is assumed that a time series $\{y_t\}$ is a realization of a deterministic trend and a stochastic component,

$$y_t = TD_t + z_t, \tag{3.1}$$

where TD_t assigns a deterministic trend, $TD_t = \beta_1 + \beta_2 t$, and z_t represents the stochastic component $\phi(L)z_t = \theta(L)\varepsilon_t$ with $\varepsilon_t \sim$ i.i.d.; *i.e.*, an autoregressive moving average process. We distinguish two cases. First, if all roots of the autoregressive polynomial lie outside the unit circle (see Equation (1.14)), then $\{y_t\}$ is stationary around a deterministic trend. In this instance, one could remove the trend from the original series $\{y_t\}$ and fit an ARMA(p, q) to the residuals.[1]

This *trend-stationary* model is also termed an integrated model of order zero, or more compactly, the $I(0)$-model. Second, assume now that one root of the autoregressive polynomial lies on the unit circle and the remaining ones are all outside. Here, $\Delta z_t = (1 - L)z_t$ is stationary around a constant mean. The series $\{y_t\}$ is *difference-stationary* because one has to apply the first difference filter with respect to time to obtain a stationary process. As in

---

[1] A deterministic trend is most easily subtracted from a series (*i.e.*, a vector y) by issuing the following command: `detrended <- residuals(lm(y ~ seq(along = y)))`.

the trend-stationary model, this difference-stationary model is referred to as an integrated model of order one, or the $I(1)$-model for short. The meaning of "integrated" should now be obvious: Once the series has been differenced to obtain a stationary process, it must be integrated once (*i.e.*, the reversal) to achieve the original series, hence the $I(1)$-model. An ARMA(p, q)-model could then be fitted to the differenced series. This model class is termed the *autoregressive integrated moving average* (ARIMA)(p, d, q), where d refers to the order of integration; *i.e.*, how many times the original series must be differenced until a stationary one is obtained. It should be noted that unit roots (*i.e.*, roots of the autoregressive polynomial that lie on the unit circle) refer solely to the stochastic component in Equation (3.1).

The distinction between a trend- and a difference-stationary process is illustrated by the two processes

$$y_t = y_{t-1} + \mu = y_0 + \mu t, \tag{3.2a}$$

$$y_t = y_{t-1} + \varepsilon_t = y_0 + \sum_{s=1}^{t} \varepsilon_s, \tag{3.2b}$$

where μ is a fixed constant and ε_t is a white noise process. In Equation (3.2a), $\{y_t\}$ is represented by a deterministic trend, whereas in Equation (3.2b) the series is explained by its cumulated shocks (*i.e.*, a stochastic trend).

So far, the stochastic component z_t has been modeled as an ARIMA(p, d, q)-model. To foster the understanding of unit roots, we will decompose the stochastic component into a cyclical component c_t and a stochastic trend TS_t. It is assumed that the cyclical component is a mean-stationary process, whereas all random shocks are captured by the stochastic component. Now, the data-generating process for $\{y_t\}$ is decomposed into a *deterministic trend*, a *stochastic trend*, and a *cyclical component*. For the trend-stationary model, the stochastic trend is zero and the cyclical component is equal to the ARMA(p, q)-model: $\phi(L)z_t = \theta(l)\varepsilon_t$. In the case of a difference-stationary model, the autoregressive polynomial contains a unit root that can be factored out, $\phi(L) = (1-L)\phi^*(L)$, where the roots of the polynomial $\phi^*(L)$ are outside the unit circle. It is then possible to express Δz_t as a moving average process (for comparison, see Equations (1.33a) and (1.33b)):

$$\phi^*(L)\Delta z_t = \theta(L)\varepsilon_t, \tag{3.3a}$$

$$\Delta z_t = \phi^*(L)\theta(L)\varepsilon_t, \tag{3.3b}$$

$$\Delta z_t = \psi(L)\varepsilon_t. \tag{3.3c}$$

Beveridge and Nelson [1981] have shown that Equation (3.3c) can be transformed to

$$z_t = TS_t + c_t = \psi(1)S_t + \psi^*(L)\varepsilon_t, \tag{3.4}$$

where the sum of the moving average coefficients is denoted by $\psi(1)$, S_t is the sum of the past and present random shocks, $\sum_{s=1}^{t} \varepsilon_s$, and the polynomial $\psi^*(L)$ is equal to $(1-L)^{-1}[\psi(L) - \psi(1)]$ (*Beveridge-Nelson decomposition*).

The time series $\{y_t\}$ is now explained by a trend function that consists of a deterministic trend as well as a stochastic component, namely $TS_T = \psi(1)S_t$. The latter affects the absolute term in each period. Because the stochastic trend is defined as the sum of the moving average coefficients of Δz_t, it can be interpreted as the long-run impact of a shock to the level of z_t. In contrast, the cyclical component, $c_t = \psi^*(L)\varepsilon_t$, exerts no long-run impact on the level of z_t. Now, we can distinguish the following four cases: (1) $\psi(1) > 1$: the long-run impact of the shocks is greater than for the intermediate ones, and hence the series is characterized by an explosive path; (2) $\psi(1) < 1$: the impact of the shocks diminishes as time passes; (3) $\psi(1) = 0$: the time series $\{y_t\}$ is a trend-stationary process; and (4) $\psi(1) = 1$: the data-generating process is a random walk. The fourth case will be a subject in the next section.

3.2 Unit Root Processes

As stated in the last section, if the sum of the moving average coefficients $\psi(1)$ equals one, a *random walk* process results. This data-generating process has attracted much interest in the empirical literature, in particular in the field of financial econometrics. Hence, a random walk is not only a prototype for a *unit root process* but is implied by economic and financial hypotheses as well (*i.e.*, the efficient market hypothesis). Therefore, we will begin this section by analyzing random walk processes in more detail before statistical tests and strategies for detecting unit roots are presented.

A *pure random walk* without a drift is defined as

$$y_t = y_{t-1} + \varepsilon_t = y_0 + \sum_{s=1}^{t} \varepsilon_t, \tag{3.5}$$

where $\{\varepsilon_t\}$ is an i.i.d. process; *i.e.*, white noise. For the sake of simplicity, assume that the expected value of y_0 is zero and that the white noise process $\{\varepsilon_t\}$ is independent of y_0. Then it is trivial to show that (1) $E[y_t] = 0$ and $\mathrm{var}(y_t) = t\sigma^2$. Clearly, a random walk is a *non-stationary time series* process because its variance grows with time. Second, the best forecast of a random walk is its value one period earlier; *i.e.*, $\Delta y_t = \varepsilon_t$. Incidentally, it should be noted that the i.i.d. assumption for the error process $\{\varepsilon_t\}$ is important with respect to the conclusions drawn above. Suppose that the data-generating process for $\{y_t\}$ is

$$y_t = y_{t-1} + \varepsilon_t, \; \varepsilon_t = \rho\varepsilon_{t-1} + \xi_t, \tag{3.6}$$

where $|\rho| < 1$ and ξ_t is a white noise process instead. Then, $\{y_t\}$ is not a random walk process, but it still has a unit root and is a first-order non-stationary process.

Let us now consider the case of a *random walk with drift*,

$$y_t = \mu + y_{t-1} + \varepsilon_t = y_0 + \mu t + \sum_{s=1}^{t} \varepsilon_t, \qquad (3.7)$$

where, as in the pure random walk process, $\{\varepsilon_t\}$ is white noise. For $\mu \neq 0$, $\{y_t\}$ contains a deterministic trend with drift parameter μ. The sign of this drift parameter causes the series to wander upward if positive and downward if negative, whereas the size of the absolute value affects the steepness.

R Code 3.1 Stochastic and deterministic trends

```
set.seed(123456)                                               1
e <- rnorm(500)                                                2
## pure random walk                                            3
rw.nd <- cumsum(e)                                             4
## trend                                                       5
trd <- 1:500                                                   6
## random walk with drift                                      7
rw.wd <- 0.5*trd + cumsum(e)                                   8
## deterministic trend and noise                               9
dt <- e + 0.5*trd                                             10
## plotting                                                   11
par(mar=rep(5,4))                                             12
plot.ts(dt, lty=1, ylab='', xlab='')                         13
lines(rw.wd, lty=2)                                          14
par(new=T)                                                   15
plot.ts(rw.nd, lty=3, axes=FALSE)                           16
axis(4, pretty(range(rw.nd)))                               17
lines(rw.nd, lty=3)                                         18
legend(10, 18.7, legend=c('det. trend + noise (ls)',       19
                'rw drift (ls)', 'rw (rs)'),               20
        lty=c(1, 2, 3))                                    21
```

In R code 3.1, three time series have been generated. For a better comparability between them, all series have been calculated with the same sequence of random numbers drawn from a standard normal distribution. First, a pure random walk has been generated by calculating the cumulated sum of 500 random numbers stored in the vector object e. A deterministic trend has been set with the short form of the seq() function; *i.e.*, the colon operator. As a second time series model, a random walk with drift can now be easily created according to Equation (3.7). Last, the deterministic trend has been overlaid with the stationary series of normally distributed errors. All three series are plotted in Figure 3.1. By ocular econometrics, it should be evident that the statistical discrimination between a deterministic trend contaminated with noise and a random walk with drift is not easy. Likewise, it is difficult to distinguish between a random walk process and a stable AR(1)-process in which the autoregressive coefficient is close to unity. The latter two time series processes are displayed in Figure 3.2.

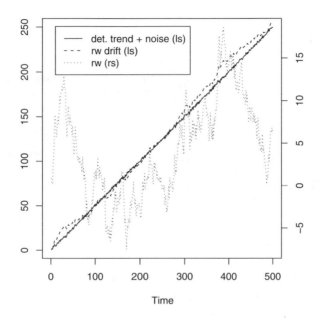

Fig. 3.1. Time series plot of deterministic and stochastic trends

Before a testing procedure for the underlying data-generating process is outlined, we will introduce a formal definition of integrated series and briefly touch on the concept of *seasonal integration*, which will be presented in more detail in Section 6.2.

In the seminal paper by Engle and Granger [1987], an *integrated series* is defined as follows.

Definition 3.1. *A series with no deterministic component that has a stationary, invertible ARMA representation after differencing d times is said to be integrated of order* d, *which is denoted as* $x_t \sim I(d)$.

That is, a stationary series is simply written as an $I(0)$-process, whereas a random walk is said to follow an $I(1)$-process because it has to be differenced once before stationarity is achieved. It should be noted at this point that some macroeconomic series are already differenced. For example, the real net investment in an economy is the difference of its capital stock. If investment is an $I(1)$-process, then the capital stock must behave like an $I(2)$-process.

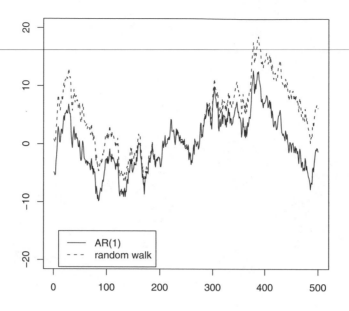

Fig. 3.2. Plot of a random walk and a stable AR(1)-process, $\phi = 0.99$

Similarly, if the inflation rate, measured as the difference of the logarithmic price index, is integrated of order one, then the price index follows an $I(2)$-process. Therefore, stationarity of $y_t \sim I(2)$ is achieved by taking the first differences of the first differences:

$$\Delta\Delta y_t = \Delta(y_t - y_{t-1}) = (y_t - y_{t-1}) - (y_{t-1} - y_{t-2}) = y_t - 2y_{t-1} + y_{t-2}. \quad (3.8)$$

If a series is already stationary $I(0)$, then no further differencing is necessary. When a series $\{y_t\}$ is a linear combination of $x_{1t} \sim I(0)$ and $x_{2t} \sim I(1)$, then $\{y_t\}$ will be an $I(1)$-process. Likewise, a linear transformation of an $I(d)$-process conserves the order of integration, $y_t \sim I(d)$, so it will be $\alpha + \beta y_t \sim I(d)$, where α and β are constants.

Until now, we have only considered data-generating processes in which the unit root occurs for its own values lagged by one period. One can generalize these processes to

$$y_t = y_{t-s} + \varepsilon_t, \quad (3.9)$$

where $s \geq 1$. If s equals a seasonal frequency of the series, then $\{y_t\}$ is determined by its prior seasonal values plus noise. As for the concept of a stochastic trend, this data-generating process is termed *stochastic seasonality*. In practice, seasonality is often accounted for by the inclusion of seasonal dummy variables or the use of seasonally adjusted data. However, there might be instances where allowing a seasonal component to drift over time is necessary. Analogously to the presentation of the unit root processes at the zero frequency, we can define the lag operator for *seasonal unit roots* as

$$\Delta_s = (1 - L^s) \tag{3.10a}$$

$$= (1 - L)(1 + L + L^2 + \ldots + L^{s-1}) = \Delta S(L). \tag{3.10b}$$

In Equation (3.10b), the unit root at the zero frequency has been factored out. Hence, a seasonally integrated series can be represented as the product of the first difference operator and the moving average seasonal filter $S(L)$. According to Engle, Granger and Hallman [1988], a seasonally integrated series can be defined as follows.

Definition 3.2. *A variable $\{y_t\}$ is said to be seasonally integrated of orders d and D, which are denoted as SI(d, D), if $\Delta^d S(L)^D y_t$ is stationary.*

Therefore, if a quarterly series $\Delta_4 y_t$ is stationary, then $\{y_t\}$ is $SI(1,1)$. Testing for seasonal unit roots is similar, although a bit more complicated than testing for unit roots at the zero frequency, which will be presented in the following paragraphs. Probably the simplest test was proposed by Hasza and Fuller [1982] and Dickey, Hasza and Fuller [1984], and a modification of it by Osborn, Chui, Smith and Birchenhall [1988]. However, a more complicated testing procedure that allows for cyclical movements at different frequencies was introduced into the literature by Hylleberg, Engle, Granger and Yoo [1990]. In R, seasonal unit root tests are implemented in the CRAN-package **uroot** (see López-de Lacalle and Díaz-Emparanza [2004]).

Recall the decomposition of a time series $\{y_t\}$ as in Equation (3.1). Now we want to investigate if the process $\{z_t\}$ does contain a unit root,

$$z_t = y_t - TD_t. \tag{3.11}$$

Hence, a deterministic trend is removed from the original series first and the residuals are tested for a unit root. Dickey and Fuller [1979] proposed the following test regression that is delineated from an assumed AR(1)-process of $\{z_t\}$ (the DF test):

$$z_t = \theta z_{t-1} + \varepsilon_t, \tag{3.12a}$$

$$z_t - z_{t-1} = \theta z_{t-1} - z_{t-1} + \varepsilon_t, \tag{3.12b}$$

$$\Delta z_t = (\theta - 1) z_{t-1} + \varepsilon_t, \tag{3.12c}$$

$$\Delta z_t = \pi z_{t-1} + \varepsilon_t. \tag{3.12d}$$

Under the null hypothesis of a unit root, $\pi = 0$, which is equivalent to $\theta = 1$ and the alternative is a trend stationary process; *i.e.*, $\pi < 0$ or $\theta < 1$. Please note that an explosive path for $\{z_t\}$, $\pi > 0$, is excluded. Equation (3.12d) can be estimated by the ordinary least-squares method. The significance of π can be tested by the usual Student t ratio. However, this test statistic does not have the familiar Student t distribution. Under the null hypothesis, an $I(0)$-variable is regressed on an $I(1)$-variable in Equation (3.12d). In this case, the limiting distribution of the Student t ratio is not normal. Fortunately, critical values have been calculated by simulation and are publicized in Fuller [1976], Dickey and Fuller [1981], and MacKinnon [1991], for instance.

So far, we have only stated that a deterministic trend is removed before testing for a unit root. In reality, neither the existence nor the form of the deterministic component is known *a priori*. Hence, we have to choose from the set of deterministic variables DV_t the one that best suits the data-generating process. The most obvious candidates such as DV_t are simply a constant, a linear trend, or higher polynomials in the trend function; *i.e.*, square or cubic. In general, only the first two are considered. The aim of characterizing the noise function $\{z_t\}$ is still the same, but now we have to take the various DV_t as deterministic regressors DR_t into account, too. The two-step procedure described above (Equations (3.11) and (3.12)) can be carried out in one equation,

$$\Delta y_t = \boldsymbol{\tau}' DR_t + \pi y_{t-1} + u_t, \qquad (3.13)$$

where $\boldsymbol{\tau}$ is the coefficient vector of the deterministic part and $\{u_t\}$ assigns an error term. For the one-step procedure, a difficulty now arises because, under the validity of the null hypothesis, the deterministic trend coefficient $\boldsymbol{\tau}$ is null, whereas under the alternative it is not. Hence, the distribution of the Student t ratio of π now depends on these nuisance parameters, too. The reason for this is that the true deterministic component is unknown and must be estimated. Critical values for different deterministic components can be found in the literature cited above as well as in Ouliaris, Park and Phillips [1989].

A weakness of the original DF test is that it does not take a possible serial correlation of the error process $\{u_t\}$ into account. Dickey and Fuller [1981] have suggested replacing the AR(1)-process for $\{z_t\}$ in Equation (3.12a) with an ARMA(p, q)-process, $\phi(L)z_t = \theta(L)\varepsilon_t$. If the noise component is an AR(p)-process, it can be shown that the test regression

$$\Delta y_t = \boldsymbol{\tau}' DR_t + \pi y_{t-1} + \sum_{j=1}^{k} \gamma_j \Delta y_{t-j} + u_t \text{ with } k = p - 1 \qquad (3.14)$$

ensures that the serial correlation in the error is removed. This test regression is called the *augmented Dickey-Fuller* (ADF) test. Several methods for selecting k have been suggested in the literature. The most prominent one is the *general-to-specific* method. Here, one starts with an *a priori* chosen upper bound $k_{\max}$ and then drops the last lagged regressor if it is insignificant. In this case, the Student t distribution is applicable. You repeat these steps until the last lagged regressor is significant; otherwise you drop it each time the equation is reestimated. If no endogenously lagged regressor turns out to be significant, you choose $k = 0$; hence the DF test results. This procedure will asymptotically yield the correct or greater lag order to the true order with probability one. Other methods for selecting an appropriate order k are based on information criteria, such as Akaike [1981] (AIC) or Schwarz [1978] (SC). Alternatively, the lag order can be determined by testing the residuals for a lack of serial correlation, as can be tested via the Ljung-Box Portmanteau test (LB) or a Lagrange multiplier (LM) test. In general, the SC, LB, or LM tests coincide with respect to selecting an optimal lag length k, whereas the AIC and the general-to-specific method will mostly imply a lag length at least as large as those of the former methods.

Once the lag order k is empirically determined, the following steps involve a *testing procedure*, as illustrated graphically in Figure 3.3. First, the encompassing ADF-test equation

$$\Delta y_t = \beta_1 + \beta_2 t + \pi y_{t-1} + \sum_{j=1}^{k} \gamma_j \Delta y_{t-j} + u_t \tag{3.15}$$

is estimated. The further steps to be taken are dependent on this result until one can conclude that the series is

  i) stationary around a zero mean,
 ii) stationary around a non-zero mean,
iii) stationary around a linear trend,
 iv) contains a unit root with zero drift, or
  v) contains a unit root with non-zero drift.

To be more concrete, the testing strategy starts by testing if $\pi = 0$ using the t statistic τ_τ. This statistic is not standard Student t distributed, but critical values can be found in Fuller [1976]. If this test is rejected, then there is no need to proceed further. The testing sequence is continued by an F type test Φ_3 with $H_0 : \beta_2 = \pi = 0$ using the critical values tabulated in Dickey and Fuller [1981]. If it is significant, then test again for a unit root using the standardized normal. Otherwise, if the hypothesis $\beta_2 = 0$ cannot be rejected, reestimate Equation (3.15) but without a trend. The corresponding t and F statistics for testing if $H_0 : \pi = 0$ and $H_0 : \beta_1 = \pi = 0$ are denoted by $\tau_\mu(\tau)$ and Φ_1. Again, the critical values for these test statistics are provided in the literature cited above. If the null hypothesis of $\tau_\mu(\tau)$ is rejected, then there is again no need to go further. If it is not, then employ the F statistic Φ_1 for testing the presence of a constant and a unit root.

However, the testing procedure does not end here. If the hypothesis $\pi = 0$ cannot be rejected in Equation (3.15), then the series might be integrated of a higher order than zero. Therefore, one has to test whether the series is $I(1)$ or possibly $I(2)$, or even integrated to a higher degree. A natural approach would be to apply the DF or ADF tests to

$$\Delta \Delta y_t = \pi \Delta y_{t-1} + u_t. \tag{3.16}$$

If the null hypothesis $\pi = 0$ is rejected, then $\Delta y_t \sim I(0)$ and $y_t \sim I(1)$; otherwise one subsequently must test whether $y_t \sim I(2)$. This testing procedure is termed *bottom-up*. However, two possibilities arise from using this bottom-up approach. First, the series cannot be transformed to stationarity regardless of how many times the difference operator is applied. Second, the danger of over-differencing exists; that is, one falsely concludes an integration order higher than the true one. This can be detected by high positive values of the DF-test statistic. This risk can be circumvented by a general-to-specific testing strategy proposed by Dickey and Pantula [1987]. They recommend starting from the highest sensible order of integration, say $I(2)$, and then testing downward to the stationary case.

So far, we have only considered the DF and the ADF tests as means to detect the presence of unit roots. Since the early 1980s, numerous other statistical tests have been proposed in the literature. The most important and widely used ones will be presented in the second part of the book.

3.3 Long-Memory Processes

So far, we have considered data-generating processes that are either stationary or integrated of an integer order higher than zero (for example, the random walk as a prototype of an $I(1)$-series). Hence, it is a knife-edge decision if a series is $I(1)$ or $I(0)$ or is integrated at an even higher integer order. Furthermore, it has been shown that, for a $y_t \sim I(1)$-series, the ACF declines linearly, and for a stationary $y_t \sim I(0)$-process, the ACF declines exponentially so that observations separated by a long time span may be regarded as independent. However, some empirically observed time series share neither of these characteristics, even though they are transformed to stationarity by suitable differencing. These time series still exhibit a dependency between distant observations. They occur in many disciplines, such as finance, geophysical sciences, hydrology, and macroeconomics. Although arguing heuristically, Granger [1980] provides a theoretical justification for these processes. To cope with such time series, our current model class has to be enlarged by so-called *fractionally integrated* processes (*i.e., long-memory processes*). The literature about *fractionally integrated processes* has grown steadily since their detection in the early 1950s. Baillie [1996] cites in his survey about these processes 138 articles and 38 background references.

Before the more encompassing class of *autoregressive fractionally integrated moving average* (ARFIMA) processes is introduced, it is noteworthy to define a long-memory process and the filter for transforming fractionally integrated series.

First, we draw on the definition of McLeod and Hipel [1978].

Definition 3.3. *A process is said to possess a long memory if*

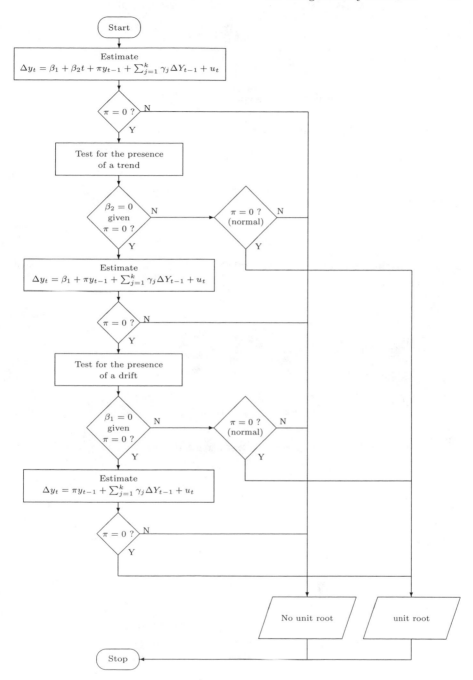

Fig. 3.3. Testing sequence for unit roots

$$\lim_{T\to\infty} \sum_{j=-T}^{T} |\rho_j| \tag{3.17}$$

is non-finite.

This is equivalent to stating that the spectral density of a long-memory process becomes unbounded at low frequencies.[2]

Second, recall that an integrated process of order d can be written as

$$(1-L)^d y_t = \psi(L)\varepsilon_t, \tag{3.18}$$

where absolute or square summability of ψ_j is given; *i.e.*, $\sum_{j=0}^{\infty} |\psi_j| < \infty$ or $\sum_{j=0}^{\infty} \psi_j^2 < \infty$. Pre-multiplying Equation (3.18) by $(1-L)^{-d}$ yields

$$y_t = (1-L)^{-d}\psi(L)\varepsilon_t. \tag{3.19}$$

Now, define the function $f(z) = (1-z)^{-d}$ for the scalar z. The derivatives of this function are

$$\frac{df}{dz} = d(1-z)^{-d-1}, \tag{3.20a}$$

$$\frac{d^2 f}{dz^2} = (d+1)d(1-z)^{-d-2}, \tag{3.20b}$$

$$\vdots$$

$$\frac{d^j f}{dz^j} = (d+j-1)(d+j-2)\cdots(d+1)d(1-z)^{-d-j}. \tag{3.20c}$$

Therefore, the fractional difference operator for $d \in (-\frac{1}{2}, \frac{1}{2}]$ can be expressed as

$$(1-L)^d = \sum_{j=0}^{\infty} \binom{d}{j}(-1)^j L^j \tag{3.21}$$

by making use of a power series expansion around $z = 0$ and the binomial theorem. The coefficient sequence $\binom{d}{j}(-1)^j$ is square summable and can be expressed in terms of the Gamma function $\Gamma()$ as

$$\binom{d}{j}(-1)^j = \frac{\Gamma(-d+j)}{\Gamma(-d)\Gamma(j+1)}. \tag{3.22}$$

---

[2] For an exposition on frequency domain analysis, the reader is referred to G. G. Judge, W. E. Griffiths, R. C. Hill, H. Lütkepohl, and T. Lee, *The Theory and Practice of Econometrics*, John Wiley and Sons, New York, 2nd ed., 1985, and P. Bloomfield, *Fourier Analysis of Time Series: An Introduction*, John Wiley and Sons, New York, 2nd ed., 2000. The *spectral density* of a series can be estimated by the function `spectrum()` in R. For more information on how this is implemented, the reader is referred to Venables and Ripley [2002] and the function's documentation.

Two points are worth noting. When $d > \frac{1}{2}$, an integer difference operator can be applied first. Incidentally, in this case, the process becomes non-stationary with unbounded variance. Robinson [1994] calls such a process "less non-stationary" than a unit root process, smoothly bridging the gulf between $I(0)$- and $I(1)$-processes. Second, because in practice no series with infinite observations are at hand, one truncates the expression in Equation (3.21) for values y_{t-j} outside the sample range and sets $y_{t-j} = 0$,

$$y_t^* = \sum_{j=0}^{\infty} \frac{\Gamma(-d+j)}{\Gamma(-d)\Gamma(j+1)} \, y_{t-j}, \tag{3.23}$$

where y_t^* assigns the fractional differenced series.

The now to be introduced ARFIMA(p, d, q) class was developed independently by Granger and Joyeux [1980] and Hosking [1981]. The estimation and simulation of these models is implemented in R within the contributed package `fracdiff` (see Fraley, Leisch and Maechler [2004]). Formally, an ARFIMA(p, d, q)-model is defined as follows.

Definition 3.4. *The series $\{y_t\}$ is an invertible and stationary ARFIMA(p, d, q)-process if it can be written as*

$$\Delta^d y_t = z_t, \tag{3.24}$$

where $\{z_t\}_{t=-\infty}^{\infty}$ is an ARMA(p, q)-process such that $z_t = \phi_p(L)^{-1}\theta_q(L)\varepsilon_t$ and both lag polynomials have their roots outside the unit circle, where ε_t is a zero-mean i.i.d. random variable with variance σ^2 and $d \in (-0.5, 0.5]$.

For parameter values $0 < d < 0.5$, the process is long-memory, and for the range $-0.5 < d < 0$, the sum of absolute values of its autocorrelations tends to a constant. In this case, the process exhibits negative dependency between distant observations and is therefore termed "*anti-persistent*" or to have "*intermediate memory.*" Regardless of whether the process $\{y_t\}$ is long-memory or intermediate memory, as long as $d > -0.5$, it has an invertible moving average representation. How is the long-memory behavior incorporated in such a process? It can be shown that the *autocorrelation function* (ACF) of long-memory processes declines hyperbolically instead of exponentially as would be the case for stable ARMA(p, q)-models. The speed of the decay depends on the parameter value d. For instance, given a fractional white noise process ARFIMA(0, d, 0), Granger and Joyeux [1980] and Hosking [1981] have proved that the autocorrelations are given by

$$\rho_j = \frac{\Gamma(j+d)\Gamma(1-d)}{\Gamma(j-d+1)\Gamma(d)}. \tag{3.25}$$

The counterpart of this behavior in the frequency domain analysis is an unbounded spectral density as the frequency ω tends to zero. In R code 3.2, an

ARIMA(0.4, 0.0, 0.0) and an ARFIMA(0.4, 0.4, 0.0) have been generated, and their ACF as well as *spectral densities* are displayed in Figure 3.4.

R Code 3.2 ARMA versus ARFIMA model

```
library(fracdiff)                                                    1
set.seed(123456)                                                     2
# ARFIMA(0.4,0.4,0.0)                                                3
y1 <- fracdiff.sim(n=1000, ar=0.4, ma=0.0, d=0.4)                    4
# ARIMA(0.4,0.0,0.0)                                                 5
y2 <- arima.sim(model=list(ar=0.4), n=1000)                          6
# Graphics                                                           7
op <- par(no.readonly=TRUE)                                          8
layout(matrix(1:6, 3, 2, byrow=FALSE))                               9
plot.ts(y1$series,                                                   10
        main='Time series plot of long memory',                     11
        ylab='')                                                     12
acf(y1$series, lag.max=100,                                          13
    main='Autocorrelations of long memory')                         14
spectrum(y1$series,                                                  15
         main='Spectral density of long memory')                    16
plot.ts(y2,                                                          17
        main='Time series plot of short memory', ylab='')           18
acf(y2, lag.max=100,                                                 19
    main='Autocorrelations of short memory')                        20
spectrum(y2, main='Spectral density of short memory')               21
par(op)                                                              22
```

A long-memory series with 1000 observations has been generated with the function `fracdiff.sim()` contained in the package **fracdiff**, whereas the short memory series has been calculated with the function `arima.sim()` (see command lines 4 and 6).[3] As can be clearly seen in Figure 3.4, the autocorrelations decline much more slowly compared with the stationary AR(1)-model, and its spectral density is higher by about a factor of 100 as $\omega \to 0$.

Up to now, the question of how to estimate the fractional difference parameter d or to detect the presence of long-memory behavior in a time series has been unanswered. We will now present three approaches to do so, where the last one deals with the simultaneous estimation of all parameters in an ARFIMA(p, d, q)-model.

The classic approach for detecting the presence of long-term memory can be found in Hurst [1951]. He proposed the rescaled range statistic, or for short the R/S *statistic*. This descriptive measure is defined as

---

[3] Functions for generating and modeling long-memory series can also be found in the contributed CRAN package **fArma** (see Würtz [2007*a*]).

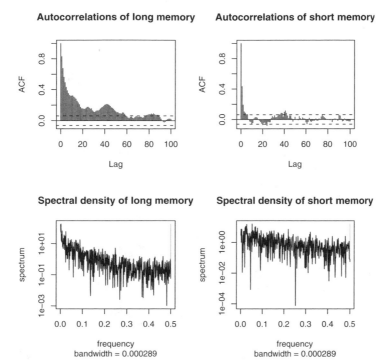

Fig. 3.4. Graphical display: ARIMA versus ARFIMA

$$R/S = \frac{1}{s_T} \left[\max_{1 \le k \le T} \sum_{j=1}^{k} (y_j - \bar{y}) - \min_{1 \le k \le T} \sum_{j=1}^{k} (y_j - \bar{y}) \right], \qquad (3.26)$$

where s_T is the usual maximum likelihood standard deviation estimator, $s_T = [\frac{1}{T} \sum_{j=1}^{T} (y_j - \bar{y})^2]^{\frac{1}{2}}$. This measure is always non-negative because the deviations from the sample mean $\bar{y}$ sum up to zero. Hence, the maximum of the partial sums will always be positive, and likewise the minimum will always be negative. Hurst [1951] showed that the probability limit

$$\plim_{T \to \infty} \left\{ T^{-H} \left(\frac{R/S}{s_t} \right) \right\} = \text{const.} \qquad (3.27)$$

is a constant and H assigns the *Hurst coefficient*. The Hurst coefficient is then estimated as

$$\hat{H} = \frac{\log(R/S)}{\log(T)}. \qquad (3.28)$$

A short-memory process is associated with a value of $H = \frac{1}{2}$, and estimated values greater than $\frac{1}{2}$ are taken as hindsight for long-memory behavior. Therefore, the differencing parameter d can be estimated as $\hat{d} = \hat{H} - \frac{1}{2}$. The R/S statistic can fairly easily be calculated in R, as shown in R code 3.3.

R **Code 3.3** R/S statistic

```
library(fracdiff)                                          1
set.seed(123456)                                           2
# ARFIMA(0.0,0.3,0.0)                                       3
y <- fracdiff.sim(n=1000, ar=0.0, ma=0.0, d=0.3)           4
# Get the data series, demean this if necessary            5
y.dm <- y$series                                           6
max.y <- max(cumsum(y.dm))                                 7
min.y <- min(cumsum(y.dm))                                 8
sd.y <- sd(y$series)                                       9
RS <- (max.y - min.y)/sd.y                                10
H <- log(RS)/log(1000)                                    11
d <- H - 0.5                                              12
```

Because the default mean in the function `fracdiff` is zero, no demeaning has to be done. The estimated Hurst coefficient is 0.7821, which implies an estimated value for d of 0.2821 that is close to its simulated value of 0.3.

Since the seminal paper of Hurst, the rescaled range statistic has received further intensive research.[4] Although it is long established that the R/S statistic has the ability to detect long-range dependence, it is, however, sensitive to short-range dependence and heteroscedasticity.[5] Hence, any incompatibility between the data and the predicted behavior of the R/S statistic under the null hypothesis of no long-run dependence need not come from long-term memory but may merely be a symptom of short-term autocorrelation. Lo [1991] proposes a modified rescaled range statistic to cope with this deficiency. The modified R/S_{mod} is defined as

$$R/S_{mod} = \frac{1}{s_T(q)} \left[\max_{1 \le k \le T} \sum_{j=1}^{k} (y_j - \bar{y}) - \min_{1 \le k \le T} \sum_{j=1}^{k} (y_j - \bar{y}) \right], \qquad (3.29)$$

where

$$s_T(q) = s_T + 2 \sum_{j=1}^{q} \omega_j(q) \hat{\gamma}_j \ , \omega_j(q) = 1 - \frac{j}{q+1} \text{ with } q < T. \qquad (3.30)$$

The maximum-likelihood standard deviation estimator is denoted by s_T and the jth-order sample autocorrelation by $\hat{\gamma}_j$. The sample autocorrelations are

---

[4] For instance, see Mandelbrot and Wallis [1968], Mandelbrot and Wallis [1969], and Davies and Harte [1987], who discuss alternative methods for estimating H. Anis and Lloyd [1976] determine the small-sample bias.

[5] For instance, see Mandelbrot [1972], Mandelbrot [1975], Mandelbrot and Wallis [1968], Davies and Harte [1987], Aydogan and Booth [1988], and Lo [1991].

weighted by the function $\omega_j(q)$ proposed in Newey and West [1987]. However, the choice of an appropriate order q is an unresolved issue.

A popular method for estimating d was proposed by Geweke and Porter-Hudak [1983]. They suggested a semi-parametric estimator of d in the frequency domain. They consider as a data-generating process $(1 - L)^d y_t = z_t$, where $z_t \sim I(0)$. This process can be represented in the frequency domain

$$f_y(\omega) = 1 - \exp(-i\omega)|^{-2d} f_z(\omega), \qquad (3.31)$$

where $f(\omega)_y$ and $f(\omega)_z$ assign the spectral densities of y_t and z_t, respectively. Equation (3.31) can be transformed to

$$\log\{f_y(\omega)\} = \left\{4\sin^2\left(\frac{\omega}{2}\right)\right\}^{-d} + \log\{f_z(\omega)\}, \qquad (3.32a)$$

$$\log\{f_y(\omega_j)\} = \log\{f_z(0)\} - d\log\left\{4\sin^2\left(\frac{\omega_j}{2}\right)\right\} + \log\left\{\frac{f_u(\omega_j)}{f_z(0)}\right\}. \qquad (3.32b)$$

The test regression is then a regression of the ordinates of the log spectral density on a trigonometric function of frequencies,

$$\log\{I_y(\omega_j)\} = \beta_1 + \beta_2 \log\left\{4\sin^2\left(\frac{\omega_j}{2}\right)\right\} + \nu_j, \qquad (3.33)$$

where $\nu_j = \log\left\{\frac{f_z(\omega_j)}{f_z(0)}\right\}$ and $j = 1, \ldots, m$. The error term is assumed to be i.i.d. with zero mean and variance $\frac{\pi}{6}$. The estimated order of fractional differencing is equal to $\hat{d} = -\hat{\beta}_2$. Its significance can be tested with either the usual t ratio distributed as Student t or one can set the residual variance equal to $\frac{\pi}{6}$. An example of this method is shown in R code 3.4, where a fractionally differenced series has been generated first with $d = 0.3$.

R Code 3.4 Geweke and Porter-Hudak method

```
library ( fracdiff )                                            1
set . seed (123456)                                            2
y <- fracdiff . sim (n=1000, ar=0.0, ma=0.0, d=0.3)           3
y . spec <- spectrum(y$series , plot=FALSE)                    4
lhs <- log (y . spec$spec )                                    5
rhs <- log (4*( sin (y . spec$freq /2))^2)                     6
gph . reg <- lm(lhs ~ rhs )                                    7
gph . sum <- summary(gph . reg )                               8
sqrt (gph . sum$cov . unscaled* pi /6) [2,2]                   9
```

The results for the simulated fractionally differenced series are given in Table 3.1. The negative of the estimated coefficient $\hat{\beta}_2$ is 0.2968, which is close to its true value of $d = 0.3$ and highly significant on both accounts; *i.e.*, its t

Table 3.1. Results of Geweke and Porter-Hudak method

Variable	Estimate	Std. Error	t-value	Pr($>|t|$)
(Intercept)	−1.6173	0.1144	−14.1370	0.0000
rhs	−0.2968	0.0294	−10.1109	0.0000

value, as well as the computed standard error with residual variance equal $\frac{\pi}{6}$. Please note that a major issue with this approach is the selection of the range of frequencies to include in the regression. In R code 3.4, all frequencies have been included (*i.e.*, 500). Diebold and Rudebusch [1989] have set $m = \sqrt{T}$, and Sowell [1992] has suggested setting m to the shortest cycle associated with long-run behavior. A third possibility would be to choose m such that the estimated standard error of the regression is approximately equal to $\sqrt{\pi/6}$.

Finally, the estimation of an ARFIMA(p, d, q)-model is implemented in the contributed package **fracdiff** as function `fracdiff()`. The parameters are estimated by an approximated maximum likelihood using the method of Haslett and Raftery [1989]. To lessen the computational burden, a range for the parameter d can be supplied as the functional argument. In the case of a "less non-stationary" series (*i.e.*, $d > \frac{1}{2}$), the estimation fails and the series must be integer differenced first. In this case, the fractional differencing filter $(1 - L)^d$ is a combination of Equation (3.21) and integer differencing.

Summary

In this chapter, a more encompassing data-generating process that was introduced into the literature by Campbell and Perron [1991] has been presented. You should now be familiar with the concepts of trend- versus difference-stationary and the decomposition of a time series into a deterministic trend, a stochastic trend, and a cyclical component. Furthermore, unit root processes have been introduced as a subclass of random walk processes. How one applies a sequential testing strategy to detect the underlying data-generating process of a possible non-stationary time series was discussed. The important definitions of integrated, seasonally integrated, and fractionally integrated time series processes have been presented, too, where the latter can be viewed as a bridge between stationary and unit root processes, thereby closing the circle of the exposition in the first two chapters.

So far, we have addressed univariate time series analysis and multivariate analysis in the context of stationary VAR and SVAR models only. The obstacles and solutions for multivariate models with non-stationary data are the subject of the next and last chapter of Part I.

Exercises

1. Write a function in R that returns the critical values given in Fuller [1976]. Function arguments should include the test type, the significance level, and the sample size.
2. Write a function in R that implements the ADF-test regression as shown in Equation (3.14). The series; the inclusion of a constant, trend, both, or none; and the order of lagged differenced series should be included as functional arguments. The function should return a summary object of class lm.
3. Now include the function of Exercise 1 in the function of Exercise 2 such that the relevant critical values are returned aside from the summary object of class lm.
4. Generate various long and intermediate processes for different values of d and AR(p) and MA(q) orders and analyze their autocorrelation functions.
5. Write a function that estimates the Hurst coefficient (*i.e.*, the R/S statistic) and its modified version by Lo [1991] and the order of the difference operator d.
6. Write a function for the single-equation estimation of d as proposed by Geweke and Porter-Hudak [1983].
7. Apply the functions of Exercises 5 and 6 to the absolute logarithmic returns of the stock indices contained in the data set EuStockMarkets. Can you detect long-memory behavior in any of these series?

4

Cointegration

In the previous chapters, a brief explanation of univari-
ate and multivariate time series models and their char-
acteristics was presented. The focus of this chapter is
on the simultaneous modeling of time series and infer-
ences of the relationships between them if some or all
of them are integrated processes of order one. As will
be shown, the degree of integration and a careful exam-
ination of the data-generating processes are of utmost
importance. We will begin by briefly reviewing the case
of a spurious regression before we proceed by providing a
definition of cointegration and its error-correction rep-
resentation. In the last section, the more encompassing
vector error-correction models are presented.

4.1 Spurious Regression

Regression analysis plays a pivotal role in applied economics. It is widely used
to test the validity of economic theories. Furthermore, the *classic linear regres-*
sion models as in Equation (4.1) form the basis of macroeconomic forecasting
and simulation models.

$$y_t = \beta_1 x_{t,1} + \beta_2 x_{t,2} + \ldots + \beta_K x_{t,K} + \varepsilon_t , \quad \text{for } t = 1, \ldots, T, \qquad (4.1)$$

where y_t assigns the endogenous variable (*i.e.*, the regressand), the exoge-
nous variables (*i.e.*, the regressors) are included in the row vector $x_t' = (x_{t,1}, x_{t,2}, \ldots, x_{t,K})$, and ε_t is a white noise random error. One important
assumption of this model class is the stationarity of the variables; that is,

$$\lim_{T \to \infty} X'X = \mathfrak{M} \text{ and } \exists \, \mathfrak{M}^{-1}. \qquad (4.2)$$

The product moment matrix of the regressors converges to the fixed and
invertible matrix $\mathfrak{M}$. This assumption is employed, for example, in the consis-
tency proof of the ordinary least-squares (OLS) estimator. Clearly, for trending variables, as are most often encountered in the field of empirical longitudi-
nal macroeconomic data, this assumption is not met. Incidentally, if only deterministic trends are present in the data-generating processes of the variables
in question, then these can be removed before estimation of Equation (4.1)
or can be included in the regression. The inference on the coefficients is the
same regardless of which method is employed; *i.e.*, the Frisch-Waugh theorem

(see Frisch and Waugh [1933]). However, matters are different in the case of difference-stationary data. In this case, the error term is often highly corre-lated and the t and F statistics are distorted such that the null hypothesis is rejected too often for a given critical value; hence, the risk of a "*spurious regression*" or "*nonsense regression*" exists.[1] Furthermore, such regressions are characterized by a high R^2. This fact arises because the endogenous variable contains a stochastic trend and the total variation is computed as $\sum_{t=1}^{T}(y_t-\bar{y})$; *i.e.*, it is erroneously assumed that the series has a fixed mean. Hence, given the formula for calculating the unadjusted R^2,

$$R^2 = 1 - \frac{\sum_{t=1}^{T}\hat{\varepsilon}_t^2}{\sum_{t=1}^{T}(y_t - \bar{y})^2}, \tag{4.3}$$

the goodness-of-fit measure tends to unity as the denominator becomes very large because a large weight is placed on extreme observations on either side of the mean $\bar{y}$.

As a rule of thumb, Granger and Newbold [1974] suggested that one should be suspicious if the R^2 is greater than the *Durbin-Watson statistic* (see Durbin and Watson [1950], Durbin and Watson [1951], and Durbin and Watson [1971]). A theoretical basis for their finding was provided by Phillips [1986].

In R code 4.1, two unrelated random walk processes with drift have been generated and regressed on each other (see command line 8). The results are provided in Table 4.1.

R **Code 4.1** Spurious regression

```
library(lmtest)                          1
set.seed(123456)                         2
e1 <- rnorm(500)                         3
e2 <- rnorm(500)                         4
trd <- 1:500                             5
y1 <- 0.8*trd + cumsum(e1)               6
y2 <- 0.6*trd + cumsum(e2)               7
sr.reg <- lm(y1 ~ y2)                    8
sr.dw <- dwtest(sr.reg)$statistic        9
```

As can be seen, the coefficient of the regressor is significant, the adjusted R^2 of 0.9866 is close to one, and the Durbin-Watson statistic of 0.0172 is

---

[1] The spurious regression problem can be traced back to Yule [1926] and Hooker [1901]. For a historic background of nonsense regressions, see Hendry [2004] and Hendry [1986]. Hendry [1980] has provided a pretty famous example of how easy it is to create a spurious regression by regressing the logarithm of the consumer price level on the cumulative rainfall in the United Kingdom.

Table 4.1. Results of spurious regression

Variable	Estimate	Std. Error	t-value	$\Pr(>|t|)$
(Intercept)	−29.3270	1.3672	−21.4511	0.0000
y2	1.4408	0.0075	191.6175	0.0000

close to zero, as expected. For the sake of completeness, the Durbin-Watson statistic implemented in the contributed package **lmtest** has been used (see Zeileis and Hothorn [2002]). An alternative is the `durbin.watson()` function in the contributed package **car** (see Fox [2007]).

From a statistical point of view, the spurious regression problem could be circumvented by taking first differences of the $I(1)$-variables in the regression equation and using these instead. However, by applying this procedure, two new problems are incurred. First, differencing greatly attenuates large positive residual autocorrelation; hence, false inferences upon the coefficients in the regression equation could be drawn. Second, most economic theories are expressed in levels, and the implications of the long-run relationships between variables are deduced. Therefore, being obliged to use regression approaches with differenced variables would be a great obstacle in the testing of economic theories. Other means of transforming non-stationary data into stationary ones (*e.g.*, by building logarithmic ratios) have been pursued with success; for example, by Sargan [1964] and Hendry and Anderson [1977]. The reason why such a transformation is suitable in achieving stationarity is that the non-stationarities are "canceling" each other out, although this must not be true in all cases and all circumstances. All in all, a new approach is called for to deal with trending variables in the context of regression analysis.

4.2 Concept of Cointegration and Error-Correction Models

In 1981, Granger [1981] introduced the concept of *cointegration* into the literature, and the general case was publicized by Engle and Granger [1987] in their seminal paper. The idea behind cointegration is to find a linear combination between two $I(d)$-variables that yields a variable with a lower order of integration. More formally, cointegration is defined as follows.

Definition 4.1. *The components of the vector $\boldsymbol{x}_t$ are said to be cointegrated of order d, b, denoted $\boldsymbol{x}_t \sim CI(d,b)$, if (a) all components of $\boldsymbol{x}_t$ are $I(d)$ and (b) a vector $\boldsymbol{\alpha}(\neq 0)$ exists so that $z_t = \boldsymbol{\alpha}'\boldsymbol{x}_t \sim I(d-b)$, $b > 0$. The vector $\boldsymbol{\alpha}$ is called the cointegrating vector.*

The great interest in this path-breaking development among economists is mostly explained by the fact that it is now possible to detect stable long-run relationships among non-stationary variables. Consider the case of $d = 1$, $b = 1$; *i.e.*, the components in the vector $\boldsymbol{x}_t$ are all integrated of order one, but if a linear combination $\boldsymbol{\alpha}$ of these exists, then the resultant series z_t is stationary. Although the individual series are non-stationary, they are tied to each other by the cointegrating vector. In the parlance of economics, deviations from a long-run equilibrium path are possible, but these errors are characterized by a mean reversion to its stable long-run equilibrium.

Now, the question is how to estimate the cointegrating vector $\boldsymbol{\alpha}$ and how to model the dynamic behavior of $I(d)$-variables in general and for exposition purposes of $I(1)$-variables in particular?

Engle and Granger [1987] proposed a *two-step* estimation technique to do so. In the first step, a regression of the variables in the set of $I(1)$ is run,

$$y_t = \alpha_1 x_{t,1} + \alpha_2 x_{t,2} + \ldots + \alpha_K x_{t,K} + z_t \text{ for } t = 1, \ldots, T, \qquad (4.4)$$

where z_t assigns the error term. The estimated $(K + 1)$ cointegrating vector $\hat{\boldsymbol{\alpha}}$ is given by $\hat{\boldsymbol{\alpha}} = (1, -\hat{\boldsymbol{\alpha}}^*)'$, where $\hat{\boldsymbol{\alpha}}^* = (\hat{\alpha}_1, \ldots, \hat{\alpha}_K)'$. Hence, the cointegrating vector is normalized to the regressand. Engle and Granger showed that in this static regression the cointegrating vector can be consistently estimated but with a finite sample bias of magnitude $O_p(T^{-1})$. Because the usual convergence rate in the $I(0)$ case is only $O_p(T^{-1/2})$, Stock [1987] termed the OLS estimation of the cointegrating vector as "superconsistent." Incidentally, although the cointegrating vector can be *superconsistently* estimated, Stock has shown that the limiting distribution is non-normal; hence, as in the case of spurious regressions, the typical t and F statistics are not applicable. However, what has been gained is first a resurrection of the applicability of the OLS method in the case of trending variables, and second the residuals from this static regression (*i.e.*, $\hat{z}_t$) are in the case of cointegration integrated of order zero. These residuals are the errors from the long-run equilibrium path of the set of $I(1)$-variables. Whether this series is stationary (*i.e.*, the variables are cointegrated) can be tested for example with the Dickey-Fuller (DF) test or the augmented Dickey-Fuller (ADF) test. Please note that now the critical values provided in Engle and Yoo [1987] or Phillips and Ouliaris [1990] have to be considered because the series $\hat{z}_t$ is an estimated one.[2] As a rough check, the so-called *cointegrating regression Durbin-Watson* (CRDW) test proposed by Sargan and Bhargava [1983] can be calculated with the null hypothesis $CRDW = 0$. The test statistic is the same as the usual Durbin-Watson test, but the prefix "cointegrating" has been added to emphasize its utilization in the context of cointegration testing. Once the null hypothesis of a unit root in the series $\hat{z}_t$ has been rejected, the second step of the two-step procedure follows. In this second step, an *error-correction model* (ECM) is specified, the

---

[2] MacKinnon [1991] has calculated critical values for the Dickey-Fuller (DF) and augmented DF (ADF) tests based on critical surface regressions. These values are readily available in the function `unitrootTable()` contained in the package **fUnitRoots** (see Würtz [2007b]).

Engle-Granger *representation theorem*. We restrict ourselves to the bivariate case first, in which two cointegrated variables y_t and x_t, each $I(1)$, are considered. In Section 4.3, systems of cointegrated variables are then presented. The general specification of an ECM is as follows:

$$\Delta y_t = \psi_0 + \gamma_1 \hat{z}_{t-1} + \sum_{i=1}^{K} \psi_{1,i} \Delta x_{t-i} + \sum_{i=1}^{L} \psi_{2,i} \Delta y_{t-i} + \varepsilon_{1,t}, \qquad (4.5a)$$

$$\Delta x_t = \xi_0 + \gamma_2 \hat{z}_{t-1} + \sum_{i=1}^{K} \xi_{1,i} \Delta y_{t-i} + \sum_{i=1}^{L} \xi_{2,i} \Delta x_{t-i} + \varepsilon_{2,t}, \qquad (4.5b)$$

where $\hat{z}_t$ is the error from the static regression in Equation (4.4), and $\varepsilon_{1,t}$ and $\varepsilon_{2,t}$ signify white noise processes. The ECM in Equation (4.5a) states that changes in y_t are explained by their own history, lagged changes of x_t, and the error from the long-run equilibrium in the previous period. The value of the coefficient γ_1 determines the speed of adjustment and should always be negative in sign. Otherwise the system would diverge from its long-run equilibrium path. Incidentally, one is not restricted to including the error from the previous period only. It can be any lagged value because Equations (4.5a) and (4.5b) are still balanced because $\hat{z}_{t-1}$ is stationary and so is $\hat{z}_{t-k}$ with $k > 1$. Furthermore, as can be concluded from these equations and the static regression, in the case of two cointegrated $I(1)$-variables, *Granger causality* must exist in at least one direction. That is, at least one variable can help forecast the other.

In addition to the empirical examples of this method exhibited in Section 7.1, an artificial one is presented in R code 4.2.[3]

R Code 4.2 Engle-Granger procedure with generated data

```
set.seed(123456)                                                      1
e1 <- rnorm(100)                                                      2
e2 <- rnorm(100)                                                      3
y1 <- cumsum(e1)                                                      4
y2 <- 0.6*y1 + e2                                                     5
lr.reg <- lm(y2 ~ y1)                                                 6
error <- residuals(lr.reg)                                            7
error.lagged <- error[-c(99, 100)]                                   8
dy1 <- diff(y1)                                                       9
dy2 <- diff(y2)                                                      10
diff.dat <- data.frame(embed(cbind(dy1, dy2), 2))                   11
colnames(diff.dat) <- c('dy1', 'dy2', 'dy1.1', 'dy2.1')            12
ecm.reg <- lm(dy2 ~ error.lagged + dy1.1 + dy2.1,                   13
              data=diff.dat)                                        14
```

---

[3] In the package's vignette of **strucchange**, an absolute consumption function for the United States is specified as an ECM.

Table 4.2. Results of Engle-Granger procedure with generated data

Variable	Estimate	Std. Error	t-value	$\Pr(>|t|)$
(Intercept)	0.0034	0.1036	0.0328	0.9739
error.lagged	−0.9688	0.1586	−6.1102	0.0000
dy1.1	0.8086	0.1120	7.2172	0.0000
dy2.1	−1.0589	0.1084	−9.7708	0.0000

First, two random walks were created, in which the latter one, y2, has been set to 0.6*y1 + e2, where e2 is a white noise process (see command lines 2 to 5). Hence, the cointegrating vector is $(1, -0.6)$. First, the long-run equation lr.reg has been estimated by OLS. Given a sample size of 100, as expected, the estimated coefficient of the regressor y1 is close to its theoretical counterpart by having a value of 0.5811. In the following command lines, the equilibrium error is stored as error, and its lagged version has been created by simply dropping the last two entries (see command line 8). Because differences and first lagged differences of y1 and y2 are generated with the commands diff() and embed(), subtracting the last two entries of the series error is equivalent with lagging the error term once in the ensuing ECM regression (see command line 13).[4] The results of the ECM are displayed in Table 4.2. To no surprise, the equilibrium error of the last period is almost completely worked off. Its coefficient is significant and close to negative one.

So far, we have restricted the exposition to the bivariate case and hence to only one cointegrating vector. However, if the dimension n of the vector $\boldsymbol{x}'_t$ is greater than two, up to $n - 1$ distinct linear combinations could exist that would each produce a series with a lower order of integration. Therefore, by applying the Engle-Granger two-step procedure presented above in a case in which $n > 2$, one estimates a single cointegrating vector only, which would represent an average of up to $n - 1$ distinct cointegrating relationships. How to cope with multiple long-run relationships is the subject of the next section.

4.3 Systems of Cointegrated Variables

Before the *vector error-correction model* (VECM) is presented, the time series decomposition in a deterministic and a stochastic component as in Equation (3.1) is extended to the multivariate case and the broader concept of cointegration is defined.

We now assume that each component of the $(K \times 1)$ vector $\boldsymbol{y}_t$, for $t = 1, \ldots, T$, can be represented as

---

[4] ECM models can be conveniently set up and estimated by utilizing the function dynlm() in the contributed package **dynlm** (see Zeileis [2006]) or with the function dyn() contained in the package **dyn** (see Grothendieck [2005]). The reader is encouraged to try out these functions as a supplemental exercise.

$$y_{i,t} = TD_{i,t} + z_{i,t} \text{ for } i = 1, \ldots, K \text{ and } t = 1, \ldots, T, \tag{4.6}$$

where $TD_{i,t}$ assigns the deterministic component of the ith variable and $z_{i,t}$ represents the stochastic component as an autoregressive moving average (ARMA) process, $\phi_i(L)z_{i,t} = \theta_i(L)\varepsilon_{i,t}$. It is further assumed that $y_{i,t}$ contains a maximum of one unit root and all remaining ones are outside the unit circle. Campbell and Perron [1991] have then defined cointegration in a broader sense as follows.

Definition 4.2. *An $(n \times 1)$ vector of variables $\boldsymbol{y}_t$ is said to be cointegrated if at least one nonzero n-element vector $\boldsymbol{\beta}_i$ exists such that $\boldsymbol{\beta}_i' \boldsymbol{y}_t$ is trend-stationary. $\boldsymbol{\beta}_i$ is called a cointegrating vector. If r such linearly independent vectors $\boldsymbol{\beta}_i(i = 1, \ldots, r)$ exist, we say that $\{\boldsymbol{y}_t\}$ is cointegrated with cointegrating rank r. We then define the $(n \times r)$ matrix of cointegrating vectors $\boldsymbol{\beta} = (\boldsymbol{\beta}_1, \ldots, \boldsymbol{\beta}_r)$. The r elements of the vector $\boldsymbol{\beta}' \boldsymbol{y}_t$ are trend-stationary, and $\boldsymbol{\beta}$ is called the cointegrating matrix.*

This definition is broader than the one by Engle and Granger (see Definition 4.1) in the sense that now it is no longer required that each individual series be integrated of the same order. For example, some or all series can be trend-stationary. If $\boldsymbol{y}_t$ contains a trend-stationary variable, then it is trivially cointegrated and the cointegrating vector is the unit vector that selects the stationary variable. On the other hand, if all series are trend-stationary, then the system is again trivially cointegrated because any linear combination of trend-stationary variables yields a trend-stationary variable. Furthermore, non-zero linear trends in the data are also included per Equation (4.6).

Reconsider now the vector autoregression model of order p

$$\boldsymbol{y}_t = \boldsymbol{\Pi}_1 \boldsymbol{y}_{t-1} + \ldots + \boldsymbol{\Pi}_K \boldsymbol{y}_{t-p} + \boldsymbol{\mu} + \boldsymbol{\Phi} \boldsymbol{D}_t + \boldsymbol{\varepsilon}_t \text{ for } t = 1, \ldots, T, \tag{4.7}$$

where $\boldsymbol{y}_t$ assigns the $(K \times 1)$ vector of series at period t, the matrices $\boldsymbol{\Pi}_i(i = 1, \ldots, p)$ are the $(K \times K)$ coefficient matrices of the lagged endogenous variables, $\boldsymbol{\mu}$ is a $(K \times 1)$ vector of constants, and $\boldsymbol{D}_t$ is a vector of non-stochastic variables such as seasonal dummies or intervention dummies. The $(K \times 1)$ error term $\boldsymbol{\varepsilon}_t$ is assumed to be i.i.d. as $\boldsymbol{\varepsilon}_t \sim \mathcal{N}(\boldsymbol{0}, \boldsymbol{\Sigma})$.

From Equation (4.7), two versions of a VECM can be delineated. In the first form, the levels of $\boldsymbol{y}_t$ enter with lag $t - p$:

$$\Delta \boldsymbol{y}_t = \boldsymbol{\Gamma}_1 \Delta \boldsymbol{y}_{t-1} + \ldots + \boldsymbol{\Gamma}_{p-1} \Delta \boldsymbol{y}_{t-p+1} + \boldsymbol{\Pi} \boldsymbol{y}_{t-p} + \boldsymbol{\mu} + \boldsymbol{\Phi} \boldsymbol{D}_t + \boldsymbol{\varepsilon}_t, \tag{4.8a}$$

$$\boldsymbol{\Gamma}_i = -(\boldsymbol{I} - \boldsymbol{\Pi}_1 - \ldots - \boldsymbol{\Pi}_i), \text{ for } i = 1, \ldots, p-1, \tag{4.8b}$$

$$\boldsymbol{\Pi} = -(\boldsymbol{I} - \boldsymbol{\Pi}_1 - \cdots - \boldsymbol{\Pi}_p), \tag{4.8c}$$

where $\boldsymbol{I}$ is the $(K \times K)$ identity matrix. As can be seen from Equation (4.8b), the $\boldsymbol{\Gamma}_i(i = 1, \ldots, p-1)$ matrices contain the cumulative long-run impacts; hence, this specification is termed the *long-run form*. Please note that the levels of $\boldsymbol{y}_t$ enter with lag $t - p$.

The other VECM specification is of the form

$$\Delta y_t = \Gamma_1 \Delta y_{t-1} + \ldots + \Gamma_{p-1} \Delta y_{t-p+1} + \Pi y_{t-1} + \mu + \Phi D_t + \varepsilon_t, \quad (4.9a)$$

$$\Gamma_i = -(\Pi_{i+1} + \ldots + \Pi_p), \text{ for } i = 1, \ldots, p-1, \quad (4.9b)$$

$$\Pi = -(I - \Pi_1 - \cdots - \Pi_p). \quad (4.9c)$$

The Π matrix is the same as in the first specification. However, the Γ_i matrices now differ in the sense that they measure transitory effects; hence, this form of the VECM is termed the *transitory form*. Furthermore, the levels of the components in y_t enter lagged by one period. Incidentally, as will become evident, inferences drawn on Π will be the same regardless of which specification is chosen, and the explanatory power is the same.

Per assumption, the individual components of y_t are at most $I(1)$-variables (see Definition 4.2). Therefore, the left-hand side of the VECM is stationary. Besides lagged differences of y_t, the error-correction term Πy_{t-p} or, depending on the specification of the VECM, Πy_{t-1} appears. This term must be stationary, too; otherwise the VECM will not balance. The question now is, what kind of conditions must be given for the matrix Π such that the right-hand side is stationary? Three cases must be considered,

(i) $rk(\Pi) = K$,
(ii) $rk(\Pi) = 0$,
(iii) $0 < rk(\Pi) = r < K$,

where $rk()$ assigns the rank of a matrix. In the first case, all K linearly independent combinations must be stationary. This can only be the case if the deviations of y_t around the deterministic components are stationary. Equations (4.8) and (4.9) represent a standard VAR-model in levels of y_t. In the second case, in which the rank of Π is zero, no linear combination exists to make Πy_t stationary except for the trivial solution. Hence, this case would correspond to a VAR-model in first differences. The interesting case is the third one, in which $0 < rk(\Pi) = r < K$. Because the matrix does not have full rank, two $(K \times r)$ matrices α and β exist such that $\Pi = \alpha\beta'$. Hence, $\alpha\beta' y_{t-p}$ is stationary, and therefore the matrix-vector product $\beta' y_{t-p}$ is stationary. The r linear independent columns of β are the cointegrating vectors, and the rank of Π is equal to the cointegration rank of the system y_t. That is, each column represents one long-run relationship between the individual series of y_t. However, the parameters of the matrices α and β are undefined because any non-singular matrix Ξ would yield $\alpha\Xi(\beta\Xi^{-1})' = \Pi$. It implies that only the cointegration space spanned by β can be determined. The obvious solution is to normalize one element of β to one. The elements of α determine the speed of adjustment to the long-run equilibrium. It is referred to as the *loading* or *adjustment* matrix.

Johansen [1988], Johansen [1991], and Johansen and Juselius [1990] developed maximum-likelihood estimators of these cointegration vectors for an autoregressive process as in Equations (4.7) through (4.9). A thorough and concise presentation of this approach is given in a monograph by Johansen

[1995]. Their approach uses *canonical correlation* analysis as a means to reduce the information content of T observations in the K-dimensional space to a lower-dimensional one of r cointegrating vectors. Hence, the canonical correlations determine the extent to which the multicollinearity in the data will allow such a smaller r-dimensional space. To do so, $2K$ auxiliary regressions are estimated by OLS: Δy_t is regressed on lagged differences of y_t. The residuals are termed R_{0t}. In the second set of auxiliary regressions, y_{t-p} is regressed on the same set of regressors. Here, the residuals are assigned as R_{1t}. The $2K$ residual series of these regressions are used to compute the product moment matrices as

$$\hat{S}_{ij} = \frac{1}{T} \sum_{t=1}^{T} R_{it} R'_{jt} \text{ with } i, j = 0, 1. \tag{4.10}$$

Johansen showed that the likelihood-ratio test statistic of the null hypothesis that there are at most r cointegrating vectors is

$$-2\ln(Q) = -T \sum_{i=r+1}^{n} (1 - \hat{\lambda}_i), \tag{4.11}$$

where $\hat{\lambda}_{r+1}, \ldots, \hat{\lambda}_p$ are the $n - r$ smallest eigenvalues of the equation

$$|\lambda \hat{S}_{11} - \hat{S}_{10} \hat{S}_{00}^{-1} \hat{S}_{01}| = 0. \tag{4.12}$$

For ease of computation, the $(K \times K)$ matrix $\hat{S}_{11}$ can be decomposed into the product of a non-singular $(K \times K)$ matrix C such that $\hat{S}_{11} = CC'$. Equation (4.12) would then accordingly be written as

$$|\lambda I - C^{-1} \hat{S}_{10} \hat{S}_{00}^{-1} \hat{S}_{01} C'^{-1}| = 0, \tag{4.13}$$

where I assigns the identity matrix.

Johansen [1988] has tabulated critical values for the test statistic in Equation (4.11) for various quantiles and up to five cointegration relations; *i.e.*, $r = 1, \ldots, 5$. This statistic has been named the *trace statistic*.

Besides the trace statistic, Johansen and Juselius [1990] have suggested the *maximal eigenvalue statistic* defined as

$$-2\ln(Q; r|r + 1) = -T\ln(1 - \hat{\lambda}_{r+1}) \tag{4.14}$$

for testing the existence of r versus $r + 1$ cointegration relationships. Critical values for both test statistics and different specifications with respect to the inclusion of deterministic regressors are provided in the appendix of Johansen and Juselius [1990].

Once the cointegration rank r has been determined, the cointegrating vectors can be estimated as

$$\hat{\beta} = (\hat{v}_1, \ldots, \hat{v}_r), \tag{4.15}$$

where $\hat{v}_i$ are given by $\hat{v}_i = C'^{-1} e_i$ and e_i are the eigenvectors to the corresponding eigenvalues in Equation (4.13). Equivalent to this are the first r eigenvectors of $\hat{\lambda}$ in Equation (4.12) if they are normalized such that $\hat{V}' \hat{S}_{11} \hat{V} = I$ with $\hat{V} = (\hat{v}_1, \ldots, \hat{v}_K)$.

The adjustment matrix α is estimated as

$$\hat{\alpha} = -\hat{S}_{01} \hat{\beta} (\hat{\beta}' \hat{S}_{11} \hat{\beta})^{-1} = -\hat{S}_{01} \hat{\beta}. \tag{4.16}$$

The estimator for α is dependent on the choice of the optimizing β. The estimator for the matrix Π is given as

$$\hat{\Pi} = -\hat{S}_{01} \hat{\beta} (\hat{\beta}' \hat{S}_{11} \hat{\beta})^{-1} \hat{\beta}' = -\hat{S}_{01} \hat{\beta} \hat{\beta}'. \tag{4.17}$$

Finally, the variance-covariance matrix of the K-dimensional error process ε_t is given as

$$\hat{\Sigma} = \hat{S}_{00} - \hat{S}_{01} \hat{\beta} \hat{\beta}' \hat{S}_{10} = \hat{S}_{00} - \hat{\alpha} \hat{\alpha}'. \tag{4.18}$$

The first part ends with a three-dimensional example of the Johansen procedure briefly exhibited above. The estimation and testing of the cointegration rank in a VECM is implemented in the package **urca** as ca.jo().[5] Besides these two functionalities, the testing of restrictions based on α, β, or both, as well as the validity of deterministic regressors, will be presented in Section 8.1.

In R code 4.3, a system of three variables with cointegration relations have been generated with a size of 250. The series are depicted in Figure 4.1. The VECM has been estimated with the function ca.jo() (see command line 14). The default value for the test statistic is the maximum eigenvalue test. The results are provided in Table 4.3.

---

[5] Incidentally, a graphical user interface for the package **urca** is shipped in the inst subdirectory of the package as an add-in to the graphical user interface Rcmdr by Fox [2004]. It is particularly suited for teaching purposes, as the apprentice can concentrate on the methodology and is not distracted at the beginning by the function's syntax. The package can be downloaded from CRAN and is hosted as project AICTS I on R-Forge.

Table 4.3. Cointegration rank: Maximum eigenvalue statistic

Hypothesis Statistic	10%	5%	1%
$r <= 2$	4.7170	6.50	8.18 11.65
$r <= 1$	41.6943	12.91	14.90 19.19
$r = 0$	78.1715	18.90	21.07 25.75

Table 4.4. Cointegration vectors

Variable	y1.l2	y2.l2	y3.l2
y1.l2	1.0000	1.0000	1.0000
y2.l2	-4.7324	0.2274	0.1514
y3.l2	-2.1299	-0.6657	2.3153

R Code 4.3 Johansen method with artificially generated data

```
library(urca)                                              1
set.seed(12345)                                            2
e1 <- rnorm(250, 0, 0.5)                                   3
e2 <- rnorm(250, 0, 0.5)                                   4
e3 <- rnorm(250, 0, 0.5)                                   5
u1.ar1 <- arima.sim(model = list(ar = 0.75),               6
                    innov = e1, n = 250)                   7
u2.ar1 <- arima.sim(model = list(ar = 0.3),                8
                    innov = e2, n = 250)                   9
y3 <- cumsum(e3)                                          10
y1 <- 0.8 * y3 + u1.ar1                                   11
y2 <- -0.3 * y3 + u2.ar1                                  12
y.mat <- data.frame(y1, y2, y3)                           13
vecm <- ca.jo(y.mat)                                      14
jo.results <- summary(vecm)                               15
vecm.r2 <- cajorls(vecm, r = 2)                           16
class(jo.results)                                         17
slotNames(jo.results)                                     18
```

Clearly, the hypothesis of two cointegrating vectors cannot be rejected for all significance levels. The estimated cointegrating vectors are displayed in Table 4.4. In the first column, the cointegration vector associated with the largest eigenvalue is displayed.

It should be noted, however, that these cointegrating vectors are all normalized to the first variable, and the loading matrix $\hat{\alpha}$ is adjusted accordingly. Hence, econometric identification restrictions are by now not imposed, but rather only the cointegration rank is determined. Johansen [1995] pro-

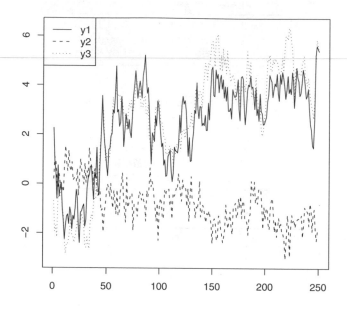

Fig. 4.1. Three $I(1)$-series with two cointegration relations

Table 4.5. Normalized cointegration vectors

Variable	ect1	ect2
y1.l2	1.0000	0.0000
y2.l2	0.0000	1.0000
y3.l2	−0.7329	0.2952

posed restricting β' in the form where the first part is an identity matrix; *i.e.*, $\beta' = [I_r : \beta_1']$, where β_1 has dimension $((K - r) \times r)$. These identifying restrictions as well as the estimation of the VECM coefficients are implemented in the function `cajorls()` contained in the package **urca**. The normalized cointegration vectors are returned in the list element `beta`, and the estimated VECM of cointegration rank r is returned as the list the element `rlm`. Both results are provided in Tables 4.5 and 4.6, respectively.

It has been shown that the VECM-model can be delineated from a VAR-model in levels. It is therefore possible to express an estimated VECM in its level-VAR form. Thereby, the estimated coefficient matrices of the VECM are retransformed according to Equations (4.8) and (4.9).

Table 4.6. VECM with $r = 2$

Variable	y1.d	y2.d	y3.d
Error-correction terms			
ect1	−0.3313	0.0646	0.0127
ect2	0.0945	−0.7094	−0.0092
Deterministic			
constant	0.1684	−0.0270	0.0253
Lagged differences			
y1.dll	−0.2277	0.0270	0.0682
y2.dll	0.1445	−0.7156	0.0405
y3.dll	0.1235	−0.2908	−0.0753

This transformation is available in the package **vars** as function vec2var().
The function's arguments are an object of formal class ca.jo and the cointegration rank r. In R code 4.4, the previously used vecm.r2 is transformed to its
level-VAR representation. The resulting object has class attribute vec2var.
For objects of this kind, the available methods and diagnostic tests outlined
in Subsections 2.2.2–2.2.6 are available, too, except for the stability() function.

R Code 4.4 VECM as VAR in levels

```
library(vars)                               1
vecm.level <- vec2var(vecm, r = 2)          2
arch.test(vecm.level)                       3
normality.test(vecm.level)                  4
serial.test(vecm.level)                     5
predict(vecm.level)                         6
irf(vecm.level, boot = FALSE)               7
fevd(vecm.level)                            8
class(vecm.level)                           9
methods(class = "vec2var")                  10
```

This section ends with some R technicalities about the package **urca**. In
this package, $S4$ classes are employed, in contrast to the older $S3$ classes. The
main difference for the user is that "slots" of an object that belong to a certain
class cannot be retrieved with object$slot as usual, but one has to use the
@ sign instead. Furthermore, the slots of an $S4$-class object cannot be shown
with names(object) as is the case with $S3$-class objects, but as shown in R
code 4.3 with the command slotNames(). The name of the class an object
belongs to can be retrieved by class(object). This brief explanation should

be sufficient for using functions in the package **urca**. Further documentation of the **methods** package can be found in R Development Core Team [2008] and Chambers [2000].

Summary

By outlining the spurious regression problem in the first section, you should have been alerted to the pitfalls when integrated time series are modeled in a multivariate context; recall the rule of thumb for detecting such nonsense regressions. In Section 4.2, the solution to this problem was presented, namely the definition of cointegration as a linear combination with a degree of lower integratedness than the two integrated processes to be investigated. In this respect, it was pointed out first that if two time series are cointegrated, then an error-correction mechanism exists and vice versa and, second, that in the case of two cointegrated $I(1)$-variables, Granger causality must exist in at least one direction. An empirical example of these issues is presented in Section 7.1. The shortcoming of the Engle-Granger procedure is that in the case of more than two integrated variables not only one cointegrating vector can exist. In fact, by applying the Engle-Granger two-step procedure in cases with more than one cointegrating vector, one would estimate a linear combination of these vectors. An encompassing definition of cointegration and the model class of VECM has been introduced to cope adequately with such instances. It has been shown that two forms of a VECM exist and that the inference with regard to the order of the space spanned by the linearly independent cointegrating vectors is the same. You should recall that neither the cointegrating vectors nor the adjustment matrix can uniquely be determined. Instead a normalization of one element of β to one is applied.

Up to now, the likelihood-based inference in cointegrated vector autoregressive models has been confined to determining the cointegration rank only. Testing various restrictions placed on the cointegration vectors and the loading matrix as well as a combination thereof are presented in Subsections 8.1.3 and 8.1.4.

Exercises

1. Write a function in R that returns the critical values for the cointegration unit root tests as given in Engle and Yoo [1987], Ouliaris et al. [1989], and MacKinnon [1991]. As functional arguments, the relevant tables, the sample size, and where applicable the number of variables in the long-run relationships should be supplied.
2. Write a function in R that implements the Engle-Granger two-step method as shown in Equations (4.4) and (4.5). The series, and the order of lagged

differenced series, should be included as functional arguments. The function should return a summary object of class lm.

3. Now include the function of Exercise 1 in the function of Exercise 2 such that the relevant critical value is returned besides a summary object of class lm.

Part II

Unit Root Tests

Testing for the Order of Integration

This chapter is the first in which the theoretical aspects laid out in Part I of the book are put into "practice." We begin by introducing the most commonly employed unit root tests in econometrics: the Dickey-Fuller test and its extensions. To discriminate between trend- and difference-stationary time series processes, a sequential testing strategy is described. Other unit root tests encountered in applied research are presented in the ensuing sections.

5.1 Dickey-Fuller Test

We now apply the augmented Dickey-Fuller (ADF) test to the data sample used by Holden and Perman [1994]. The authors applied an integration/cointegration analysis to a consumption function for the United Kingdom using quarterly data for the period 1966:Q4–1991:Q2. This data set is included in the contributed package **urca** as `Raotbl3`. The consumption series is seasonally adjusted real consumer expenditures in 1985 prices. The seasonally adjusted personal disposable income series has been deflated by the implicit consumption price index; likewise the wealth series is defined as seasonally adjusted gross personal financial wealth. All variables are expressed in their natural logarithms. Recall from Section 3.2 the test regression (3.15), which is reprinted here with the three different combinations of the deterministic part:

$$\Delta y_t = \beta_1 + \beta_2 t + \pi y_{t-1} + \sum_{j=1}^{k} \gamma_j \Delta y_{t-j} + u_{1t}, \qquad (5.1a)$$

$$\Delta y_t = \beta_1 + \pi y_{t-1} + \sum_{j=1}^{k} \gamma_j \Delta y_{t-j} + u_{2t}, \qquad (5.1b)$$

$$\Delta y_t = \pi y_{t-1} + \sum_{j=1}^{k} \gamma_j \Delta y_{t-j} + u_{3t}. \qquad (5.1c)$$

The ADF test has been implemented in the contributed packages **fUnitRoots**, **tseries**, **urca**, and **uroot** as functions `adftest()`, `adf.test()`, `ur.df()`, and `ADF.test()`, respectively. For determining the integration order as outlined in Section 3.2, we will use the function `ur.df()` for the consumption series. The reason for this is twofold. First, the three different specifications as in

Table 5.1. ADF test: Regression for consumption with constant and trend.

Variable	Estimate	Std. Error	t-value	Pr($>|t|$)
(Intercept)	0.7977	0.3548	2.2483	0.0270
z.lag.1	−0.0759	0.0339	−2.2389	0.0277
tt	0.0005	0.0002	2.2771	0.0252
z.diff.lag1	−0.1064	0.1007	−1.0568	0.2934
z.diff.lag2	0.2011	0.1012	1.9868	0.0500
z.diff.lag3	0.2999	0.1021	2.9382	0.0042

Equations (5.1a)–(5.1c) can be modeled, which is not the case for the function `adf.test()`, and second, besides the τ statistics, the F type statistics are returned in the slot `object@teststat`, with their critical values in the slot `object@cval`, as we will see shortly.

In R code 5.1, the test regressions for Models (5.1a) and (5.1b) are estimated.

R Code 5.1 ADF test: Integration order for consumption in the United Kingdom

```
library ( urca )                                                          1
data ( Raotbl3 )                                                          2
attach ( Raotbl3 )                                                        3
lc <- ts ( lc , start=c (1966,4) , end=c (1991,2) , frequency=4)         4
lc.ct <- ur.df ( lc , lags =3, type='trend ')                            5
plot ( lc.ct )                                                           6
lc.co <- ur.df ( lc , lags =3, type='drift ')                            7
lc2 <- diff ( lc )                                                       8
lc2.ct <- ur.df ( lc2 , type='trend ', lags =3)                         9
```

As a first step, a regression with a constant and a trend has been estimated (see command line 5). Three lagged endogenous variables have been included to assure a spherical error process, as is witnessed by the autocorrelations and partial autocorrelations in Figure 5.1. Including a fourth lag turns out to be insignificant, whereas specifying the test regression with only two lagged endogenous variables does not suffice to achieve serially uncorrelated errors. The summary output of this test regression is provided in Table 5.1. Next, the hypothesis $\phi_3 = (\beta_1, \beta_2, \pi) = (\beta_1, 0, 0)$ is tested by a usual F type test. That is, zero restrictions are placed on the time trend and the lagged value of `lc`. The result is displayed in Table 5.2. The test statistic has a value of 2.60. Please note that one must consult the critical values in Dickey and Fuller [1981, Table VI]. The critical value for a sample size of 100 and significance levels of 10%, 5%, and 1% are 5.47, 6.49, and 8.73, respectively. Hence, the

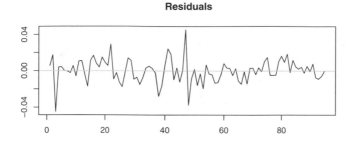

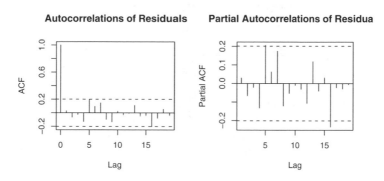

Fig. 5.1. ADF test: Diagram of fit and residual diagnostics

Table 5.2. ADF test: τ_3, ϕ_2, and ϕ_3 tests

Test Statistic		1%	5%	10%
τ_3	-2.24	-4.04	-3.45	-3.15
ϕ_2	3.74	6.50	4.88	4.16
ϕ_3	2.60	8.73	6.49	5.47

null hypothesis cannot be rejected, which implies that real consumption does contain a unit root. This finding is reiterated by a t ratio of -2.24 for the lagged endogenous variable in levels. The relevant critical values now have to be taken from Fuller [1976, Table 8.5.2], which is given for a sample size of 100 and a significance level of 5% equal to -3.45.[1]

---

[1] Instead of using the critical values in Fuller [1976], one can employ the same ones provided in Hamilton [1994] or the ones calculated by critical surface regressions in MacKinnon [1991]. In the function `ur.df()`, the critical values provided in Hamilton [1994] have been implemented for the τ statistics and the ones provided in Dickey and Fuller [1981] for the ϕ statistics. These are stored in the slot `object@cval`.

Table 5.3. ADF test: Regression for consumption with constant only

Variable	Estimate	Std. Error	t-value	$\Pr(>|t|)$
(Intercept)	0.0123	0.0851	0.1448	0.8852
z.lag.1	−0.0007	0.0079	−0.0931	0.9261
z.diff.lag1	−0.1433	0.1016	−1.4098	0.1620
z.diff.lag2	0.1615	0.1020	1.5832	0.1169
z.diff.lag3	0.2585	0.1027	2.5164	0.0136

Table 5.4. ADF test: τ_2 and ϕ_1 tests

Test Statistic	1%	5%	10%
τ_2	−0.09	−3.51	−2.89	−2.58
ϕ_1	2.88	6.70	4.71	3.86

Therefore, a unit root cannot be rejected. Next, it is tested whether the consumption series is a random walk with or without drift (see command line 7). The relevant test statistic is ϕ_2, which is provided in Table 5.2. The value of this test is 3.74 and has to be compared with the critical values provided in Dickey and Fuller [1981, Table V]. For a sample size of 100, these values are 4.16, 4.88, and 6.50 for significance levels of 10%, 5%, and 1%, respectively. The conclusion is that the consumption series behaves like a pure random walk.

One proceeds next by estimating Equation (5.1b) based on the result of the ϕ_3 test. The results are depicted in Table 5.3. For the sake of completeness, it is now tested whether in this model a drift term is absent. The test results are provided in Table 5.4. The test statistic is 2.88, which turns out to be insignificant compared with the critical values of Table IV in Dickey and Fuller [1981]. Therefore, the conclusion is that the quarterly real consumption series does contain a unit root, but neither a linear trend nor a drift is present in the data-generating process.

Finally, whether differencing the series once suffices to achieve stationarity is tested; *i.e.*, whether the series is possible $I(2)$ is tested. This test is achieved by supplying the differenced series as argument y in the function ur.df as is done in the last two command lines of R code 5.1. The results are displayed in Table 5.5. The hypothesis that the consumption series is $I(2)$ must be clearly dismissed given a t ratio of −4.39.

5.2 Phillips-Perron Test

Phillips and Perron [1988] and Perron [1988] suggested non-parametric test statistics for the null hypothesis of a unit root that explicitly allows for weak

Table 5.5. ADF test: Regression for testing $I(2)$

Variable	Estimate	Std. Error	t-value	$\Pr(>|t|)$
(Intercept)	0.0039	0.0031	1.2664	0.2087
z.lag.1	−0.8826	0.2013	−4.3853	0.0000
tt	0.0000	0.0001	0.6232	0.5348
z.diff.lag1	−0.2253	0.1873	−1.2031	0.2321
z.diff.lag2	−0.0467	0.1600	−0.2918	0.7711
z.diff.lag3	0.1775	0.1057	1.6791	0.0967

dependence and heterogeneity of the error process (the PP test). The authors consider the following two test regressions:

$$y_t = \mu + \alpha y_{t-1} + \varepsilon_t, \tag{5.2a}$$

$$y_t = \mu + \beta \left(t - \frac{1}{2}T \right) + \alpha y_{t-1} + \varepsilon_t. \tag{5.2b}$$

They define the following test statistics for Equation (5.2a):

$$Z(\hat{\alpha}) = T(\hat{\alpha} - 1) - \hat{\lambda}/\bar{m}_{yy}, \tag{5.3a}$$

$$Z(\tau_{\hat{\alpha}}) = (\hat{s}/\hat{\sigma}_{Tl})t_{\hat{\alpha}} - \hat{\lambda}'\hat{\sigma}_{Tl}/\bar{m}_{yy}^{\frac{1}{2}}, \tag{5.3b}$$

$$Z(\tau_{\hat{\mu}}) = (\hat{s}/\hat{\sigma}_{Tl})t_{\hat{\mu}} + \hat{\lambda}'\hat{\sigma}_{Tl}m_y/\bar{m}_{yy}^{\frac{1}{2}}m_{yy}^{\frac{1}{2}}, \tag{5.3c}$$

with $\bar{m}_{yy} = T^{-2}\sum(y_t - \bar{y})^2$, $m_{yy} = T^{-2}\sum y_t^2$, $m_y = T^{-3/2}\sum y_t$, and $\hat{\lambda} = 0.5(\hat{\sigma}_{Tl}^2 - \hat{s}^2)$, where $\hat{s}^2$ is the sample variance of the residuals, $\hat{\lambda}' = \hat{\lambda}/\hat{\sigma}_{Tl}^2$, and $t_{\hat{\alpha}}$, $t_{\hat{\mu}}$ are the t ratios of $\hat{\alpha}$ and $\hat{\mu}$, respectively. The long-run variance $\hat{\sigma}_{Tl}^2$ is estimated as

$$\hat{\sigma}_{Tl}^2 = T^{-1}\sum_{t=1}^{T}\hat{\varepsilon}_t^2 + 2T^{-1}\sum_{s=1}^{l}w_{sl}\sum_{t=s+1}^{T}\hat{\varepsilon}_t\hat{\varepsilon}_{t-s}, \tag{5.4}$$

where $w_{sl} = 1 - s/(l+1)$.

Similarly, the following test statistics are defined for the test regression with a linear time trend included as in Equation (5.2b):

$$Z(\tilde{\alpha}) = T(\tilde{\alpha} - 1) - \tilde{\lambda}/M, \tag{5.5a}$$

$$Z(t_{\tilde{\alpha}}) = (\tilde{s}/\tilde{\sigma}_{Tl})t_{\tilde{\alpha}} - \tilde{\lambda}'\tilde{\sigma}_{Tl}/M^{\frac{1}{2}}, \tag{5.5b}$$

$$Z(t_{\tilde{\mu}}) = (\tilde{s}/\tilde{\sigma}_{Tl})t_{\tilde{\mu}} - \tilde{\lambda}'\tilde{\sigma}_{Tl}m_y/M^{\frac{1}{2}}(M + m_y^2)^{\frac{1}{2}}, \tag{5.5c}$$

$$Z(t_{\tilde{\beta}}) = (\tilde{s}/\tilde{\sigma}_{Tl})t_{\tilde{\beta}} - \tilde{\lambda}'\tilde{\sigma}_{Tl}\left(\frac{1}{2}m_y - m_{ty}\right)/(M/12)^{\frac{1}{2}}\bar{m}_{yy}^{\frac{1}{2}}, \tag{5.5d}$$

where m_y, $\bar{m}_{yy}$, $\tilde{\lambda}$, $\tilde{\lambda}'$, and $\tilde{\sigma}_{Tl}$ are defined as in Equations (5.3a)–(5.3c) and $m_{ty} = T^{-5/2}\sum ty_t$, $t_{\tilde{\mu}}$, $t_{\tilde{\beta}}$, and $t_{\tilde{\alpha}}$ are the t ratios of $\tilde{\mu}$, $\tilde{\alpha}$, and $\tilde{\beta}$, respectively.

The scalar M is defined as $M = (1 - T^{-2})m_{yy} - 12m_{ty}^2 + 12(1 + T^{-1})m_{ty}m_y - (4 + 6T^{-1} + 2T^{-2})m_y^2$.

The critical values of these Z statistics are identical to those of the DF type tests. The advantage is that these modified tests eliminate the nuisance parameters that are present in the DF statistic if the error process does not satisfy the i.i.d. assumption. However, one problem with these tests is that it is at the researcher's discretion to choose an optimal lag number l for computing the long-run variances $\hat{\sigma}_{Tl}^2$ or $\tilde{\sigma}_{Tl}^2$ as in Equation (5.4).

The PP test is implemented as function pp.test() in the contributed package **tseries** and as function ur.pp() in the contributed package **urca**.[2] The advantage of the latter is that all test statistics are computed and returned, as well as the test regression, by applying the summary method to an object of class ur.pp. Furthermore, the lags to be included in the computation of the long-run variance can be set either manually via the argument use.lag or chosen automatically via the argument lags to be short or long, which corresponds to the integer values $4(T/100)^{\frac{1}{4}}$ and $12(T/100)^{\frac{1}{4}}$, respectively.

In R code 5.2, the PP test is applied to the consumption series used in R code 5.1.

R Code 5.2 PP test: Integration order for consumption in the United Kingdom

```
library(urca)                                        1
data(Raotbl3)                                        2
attach(Raotbl3)                                      3
lc <- ts(lc, start=c(1966,4), end=c(1991,2),         4
         frequency=4)                                5
lc.ct <- ur.pp(lc, type='Z-tau', model='trend',      6
               lags='long')                          7
lc.co <- ur.pp(lc, type='Z-tau', model='constant',   8
               lags='long')                          9
lc2 <- diff(lc)                                      10
lc2.ct <- ur.pp(lc2, type='Z-tau', model='trend',   11
                lags='long')                         12
```

First, the test is applied to Equation (5.2b), and the results are stored in the object lc.ct (see command line 6). The result of the test regression is displayed in Table 5.6. The value of the $Z(t_{\tilde{\alpha}})$ statistic is -1.92, which is insignificant. The relevant Z statistics for the deterministic regressors are 0.72 and 2.57. Both of these are insignificant if compared with the critical values of Tables II and III in Dickey and Fuller [1981] at the 5% level. These results are summarized in Table 5.7. In the next step, the trend is dropped from the

---

[2] The latter function has been ported into the package **fUnitRoots** as function urppTest().

Table 5.6. PP test: Regression for consumption with constant and trend

Variable	Estimate	Std. Error	t-value	Pr($>|t|$)
(Intercept)	0.5792	0.3622	1.5992	0.1131
y.l1	0.9469	0.0336	28.1945	0.0000
trend	0.0003	0.0002	1.6108	0.1105

Table 5.7. PP test: $Z(t_{\hat{\alpha}})$, $Z(t_{\hat{\mu}})$, and $Z(t_{\hat{\beta}})$ tests

Test	Statistic	1%	5%	10%
$Z(t_{\hat{\alpha}})$	−1.92	−4.05	−3.46	−3.15
$Z(t_{\hat{\mu}})$	0.72	3.78	3.11	2.73
$Z(t_{\hat{\beta}})$	2.57	3.53	2.79	2.38

Table 5.8. PP test: Regression for consumption with constant only

Variable	Estimate	Std. Error	t-value	Pr($>|t|$)
(Intercept)	0.0109	0.0825	0.1318	0.8954
y.l1	0.9996	0.0076	130.7793	0.0000

Table 5.9. PP test: $Z(t_{\hat{\alpha}})$ and $Z(t_{\hat{\mu}})$ tests

Test	Statistic	1%	5%	10%
$Z(t_{\hat{\alpha}})$	−0.13	−3.50	−2.89	−2.58
$Z(t_{\hat{\mu}})$	0.20	3.22	2.54	2.17

test regression and the results are stored in the object lc.co (see command line 8). The regression results are summarized in Table 5.8. Again, the null hypothesis of a unit root cannot be rejected, and the drift term is insignificant given the test statistics and critical values reported in Table 5.9. The critical values for the drift term now correspond to the ones provided for a sample size of 100 in Table I of Dickey and Fuller [1981]. So far, the conclusion about the integration order for the consumption series is the same as that obtained by applying the sequential testing procedure of the ADF test. Finally, it is checked whether differencing the series once suffices to achieve stationarity (see command lines 10 and 11). The test regression is reported in Table 5.10. The value of the test statistic $Z(t_{\hat{\alpha}})$ is −10.96, which is highly significant, and therefore it is concluded according to the results of the PP test that the consumption series behaves like a pure random walk.

Table 5.10. PP test: Regression for testing $I(2)$

Variable	Estimate	Std. Error	t-value	Pr($>|t|$)
(Intercept)	0.0073	0.0016	4.7276	0.0000
y.l1	−0.1253	0.1025	−1.2215	0.2249
trend	0.0000	0.0000	0.3606	0.7192

5.3 Elliott-Rothenberg-Stock Test

A shortcoming of the two previously introduced unit root tests is their low power if the true data-generating process is an AR(1)-process with a coefficient close to one. To improve the power of the unit root test, Elliott, Rothenberg and Stock [1996] proposed a local to unity detrending of the time series (ERS tests). The authors developed feasible point-optimal tests, denoted as P_T^μ and P_T^τ, which take serial correlation of the error term into account. The second test type is denoted as the DF-GLS test, which is a modified ADF-type test applied to the detrended data without the intercept. The following model is entertained as the data-generating process for the series $y_1, \ldots, y_T$:

$$y_t = d_t + u_t, \tag{5.6a}$$
$$u_t = au_{t-1} + v_t, \tag{5.6b}$$

where $d_t = \beta' z_t$ is a deterministic component; *i.e.*, a constant or a linear trend is included in the $(q \times 1)$ vector z_t, and v_t is a stationary zero-mean error process. In the case of $a = 1$, Equations (5.6a) and (5.6b) imply an integration order $I(1)$ for y_t, whereas $|a| < 1$ yields a stationary process for the series.

Let us first focus on the feasible point-optimal test statistic, which is defined as

$$P_T = \frac{S(a = \bar{a}) - \bar{a}S(a = 1)}{\hat{\omega}^2}, \tag{5.7}$$

where $S(a = \bar{a})$ and $S(a = 1)$ are the sums of squared residuals from a least-squares regression of y_a on Z_a with

$$y_a = (y_1, y_2 - ay_1, \ldots, y_T - ay_{T-1}), \tag{5.8a}$$
$$Z_a = (z_1, z_2 - az_1, \ldots, z_T - az_{T-1}); \tag{5.8b}$$

hence, y_a is a T-dimensional column vector and Z_a defines a $(T \times q)$ matrix. The estimator for the variance of the error process v_t can be estimated with

$$\hat{\omega} = \frac{\hat{\sigma}_\nu^2}{(1 - \sum_{i=1}^p \hat{\alpha}_i)^2}, \tag{5.9}$$

where $\hat{\sigma}_\nu^2$ and $\hat{\alpha}_i$ for $i = 1, \ldots, p$ are taken from the auxiliary ordinary least-squares (OLS) regression

$$\Delta y_t = \alpha_0 + \alpha_1 \Delta y_{t-1} + \ldots + \Delta y_{t-p} + \alpha_{p+1} + \nu_t. \tag{5.10}$$

Finally, the scalar $\bar{a}$ is set to $\bar{a} = 1 + \bar{c}/T$, where $\bar{c}$ denotes a constant. Depending on the deterministic components in z_t, $\bar{c}$ is set either to -7 in the case of a constant or to -13.5 in the case of a linear trend. These values have been derived from the asymptotic power functions and its envelope. Critical values of the P_T^μ and P_T^τ tests are provided in Table I of Elliott et al. [1996].

Next, these authors propose a modified ADF-type test, which is the t statistic for testing $\alpha_0 = 0$ in the homogeneous regression

$$\Delta y_t^d = \alpha_0 y_{t-1}^d + \alpha_1 \Delta y_{t-1}^d + \ldots + \alpha_p \Delta y_{t-p}^d + \varepsilon_t, \tag{5.11}$$

where y_t^d are the residuals in the auxiliary regression $y_t^d \equiv y_t - \hat{\beta}' z_t$. When there is no intercept, one can apply the critical values of the typical DF-type t tests; in the other instances, critical values are provided in Table I of Elliott et al. [1996].

Both test types have been implemented as function ur.ers() in the contributed package **urca**. The function allows the provision of the number of lags to be included in the test regression for the DF-GLS test via its argument lag.max. The optimal number of lags for estimating $\hat{\omega}$ is determined by the Bayesian information criterion (BIC). A summary method for objects of class ur.ers exists that returns the test regression in the case of the DF-GLS test and the value of the test statistic with the relevant critical values for the 1%, 5%, and 10% significance levels for both tests. In R code 5.3, both test types are applied to the logarithm of real gross national product (GNP) used in the seminal paper of Nelson and Plosser [1982].

R Code 5.3 ERS tests: Integration order for real GNP in the United States

```
library(urca)                                                          1
data(nporg)                                                            2
gnp <- log(na.omit(nporg[, "gnp.r"]))                                  3
gnp.d <- diff(gnp)                                                     4
gnp.ct.df <- ur.ers(gnp, type = "DF-GLS",                             5
                    model = "trend", lag.max = 4)                      6
gnp.ct.pt <- ur.ers(gnp, type = "P-test",                             7
                    model = "trend")                                   8
gnp.d.ct.df <- ur.ers(gnp.d, type = "DF-GLS",                        9
                     model = "trend", lag.max = 4)                    10
gnp.d.ct.pt <- ur.ers(gnp.d, type = "P-test",                        11
                     model = "trend")                                 12
```

First, the P_T^τ and DF-GLS$^\tau$ are applied to the series, where in the case of the DF-GLS$^\tau$, four lags have been added (see command lines 5 and 6).

Table 5.11. ERS tests: P_T^τ and DF-GLS$^\tau$ for real GNP of the United States

Test	Statistic	1%	5%	10%
P_T^τ	6.65	4.26	5.64	6.79
DF-GLS$^\tau$	-2.08	-3.58	-3.03	-2.74

Table 5.12. ERS tests: P_T^τ and DF-GLS$^\tau$ for testing $I(2)$

Test	Statistic	1%	5%	10%
P_T^τ	2.68	4.26	5.64	6.79
DF-GLS$^\tau$	-4.13	-3.58	-3.03	-2.74

The test results are displayed in Table 5.11. Both tests imply a unit root for the data-generating process. In the second step, both tests are applied to the differenced series. As the results summarized in Table 5.12 imply, differencing the series once suffices to achieve stationarity.

5.4 Schmidt-Phillips Test

In Section 5.3, unit root tests have been described that are more powerful than the usual DF-type tests. Another drawback of the DF-type tests is that the nuisance parameters (*i.e.*, the coefficients of the deterministic regressors) are either not defined or have a different interpretation under the alternative hypothesis of stationarity. To elucidate this point, consider the test regressions as in Equations (5.1a)–(5.1c) again. Equation (5.1c) allows neither a non-zero level nor a trend under both the null and the alternative hypotheses. Whereas in Equations (5.1a) and (5.1b) these regressors are taken into account, now these coefficients have a different interpretation under the null and the alternative hypotheses. The constant term β_1 in Equation (5.1b) has the interpretation of a deterministic trend under the null hypothesis (*i.e.*, $\pi = 1$), but it has to be considered as a level regressor under the alternative. Likewise, in Equation (5.1a), β_1 represents a linear trend and β_2 represents a quadratic trend under the null hypothesis of integratedness, but these coefficients have the interpretation of a level and linear trend regressor under the alternative hypothesis of stationarity. Schmidt and Phillips [1992] proposed a Lagrange multiplier (LM)-type test statistic that defines the same set of nuisance parameters under both the null and the alternative hypotheses. Furthermore, they consider higher polynomials than a linear trend. The authors consider the model

$$y_t = \alpha + \boldsymbol{Z}_t\boldsymbol{\delta} + x_t, \tag{5.12a}$$

$$x_t = \pi x_{t-1} + \varepsilon_t, \tag{5.12b}$$

where ε_t are i.i.d. $\mathcal{N}(0, \sigma^2)$ and $\boldsymbol{Z}_t = (t, t^2, \ldots, t^p)$. The test statistic $\tilde{\rho}$ is then constructed by running the regression

$$\Delta y_t = \Delta \boldsymbol{Z}_t \boldsymbol{\delta} + u_t \qquad (5.13)$$

first and calculating $\tilde{\psi}_x = y_1 - \boldsymbol{Z}_1 \tilde{\delta}$, where $\tilde{\delta}$ is the OLS estimate of δ in Equation (5.13). Next, a series $\tilde{S}_t$ is defined as $\tilde{S}_t = y_t - \tilde{\psi}_x - \boldsymbol{Z}_t \tilde{\delta}$. Finally, the test regression is then given by

$$\Delta y_t = \Delta \boldsymbol{Z}_t \gamma + \phi \tilde{S}_{t-1} + v_t, \qquad (5.14)$$

where v_t assigns an error term. The test statistic is then defined as $Z(\rho) = \frac{\tilde{\rho}}{\tilde{\omega}^2} = \frac{T\tilde{\phi}}{\tilde{\omega}^2}$ with $\tilde{\phi}$ as the OLS estimate of ϕ in Equation (5.14), and an estimator for ω^2 is given by

$$\hat{\omega}^2 = \frac{T^{-1} \sum_{i=1}^{T} \hat{\varepsilon}_t^2}{T^{-1} \sum_{i=1}^{T} \hat{\varepsilon}_t^2 + 2T^{-1} \sum_{s=1}^{l} \sum_{t=s+1}^{T} \hat{\varepsilon}_t \hat{\varepsilon}_{t-s}}, \qquad (5.15)$$

where $\hat{\varepsilon}_t$ are the residuals from Equation (5.12). Depending on the sample size and the order of the polynomial $\boldsymbol{Z}$, critical values are provided in Schmidt and Phillips [1992]. Aside from this test statistic, one can also apply the t ratio statistic $Z(\tau) = \frac{\tau}{\tilde{\omega}^2}$ for testing $\phi = 0$. As shown by a Monte Carlo simulation, these tests fare better in terms of power compared with the corresponding DF-type test statistic.

These two tests have been implemented as function `ur.sp()` in the contributed package **urca**. As arguments, the series name, the test type (either `tau` for $\tilde{\tau}$ or `rho` for $\tilde{\rho}$), the polynomial degree, and the significance level have to be entered in `ur.sp()`. In R code 5.4, these tests have been applied to the nominal GNP series of the United States expressed in millions of current U.S. dollars as used by Nelson and Plosser [1982]. By eyeball inspection of the series as displayed in Figure 5.2, a quadratic trend is assumed.

R Code 5.4 SP test: Integration order for nominal GNP of the United States

```
library(urca)                                        1
data(nporg)                                          2
gnp <- na.omit(nporg[, "gnp.n"])                     3
gnp.tau.sp <- ur.sp(gnp, type = "tau", pol.deg=2,    4
                signif=0.05)                         5
gnp.rho.sp <- ur.sp(gnp, type = "rho", pol.deg=2,    6
                signif=0.05)                         7
```

This setting is evidenced by a significant coefficient of the quadratic trend regressor in the unconstrained model (5.12a) as displayed in Table 5.13. The results of the two tests are displayed in Table 5.14. Both tests indicate an integration order of at least $I(1)$ for a significance level of 5%.

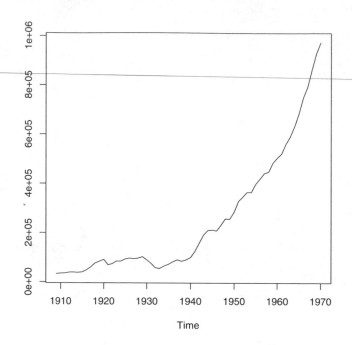

Fig. 5.2. Nominal GNP of the United States

Table 5.13. SP test: Result of level regression with polynomial of order two.

Variable	Estimate	Std. Error	t-value	$\Pr(>|t|)$
(Intercept)	9355.4016	7395.3026	1.2650	0.2110
y.lagged	0.9881	0.0411	24.0328	0.0000
trend.exp1	−982.3230	610.1348	−1.6100	0.1129
trend.exp2	30.2762	16.1573	1.8738	0.0661

Table 5.14. SP tests: $\tilde{\tau}$ and $\tilde{\rho}$ for nominal GNP of the United States

Test Statistic		1%	5%	10%
$\tilde{\tau}$		−2.03	−4.16	−3.65	−3.34
$\tilde{\rho}$		−8.51	−30.40	−23.70	−20.40

5.5 Kwiatkowski-Phillips-Schmidt-Shin Test

Kwiatkowski, Phillips, Schmidt and Shin [1992] proposed an LM test for testing trend and/or level stationarity (the KPSS test). That is, now the null
hypothesis is a stationary process, whereas in the former tests it was a unit
root process. Taking the null hypothesis as a stationary process and the unit
root as an alternative is in accordance with a conservative testing strategy.
One should always seek tests that place the hypothesis we are interested in
as the alternative one. Hence, if we then reject the null hypothesis, we can
be pretty confident that the series indeed has a unit root. Therefore, if the
results of the tests above indicate a unit root but the result of the KPSS test
indicates a stationary process, one should be cautious and opt for the latter
result. Kwiatkowski et al. consider the following model:

$$y_t = \xi t + r_t + \varepsilon_t, \tag{5.16a}$$

$$r_t = r_{t-1} + u_t, \tag{5.16b}$$

where r_t is a random walk and the error process is assumed to be i.i.d. $(0, \sigma_u^2)$.
The initial value r_0 is fixed and corresponds to the level. If $\xi = 0$, then this
model is in terms of a constant only as deterministic regressor. Under the null
hypothesis, ε_t is stationary and therefore y_t is either trend-stationary or in the
case of $\xi = 0$ level-stationary. The test statistic is constructed as follows. First,
regress y_t on a constant or on a constant and a trend, depending on whether
one wants to test level- or trend-stationarity. Then, calculate the partial sums
of the residuals $\hat{\varepsilon}_t$ from this regression as

$$S_t = \sum_{i=1}^{t} \hat{\varepsilon}_i , t = 1, 2, \ldots, T. \tag{5.17}$$

The test statistic is then defined as

$$LM = \frac{\sum_{t=1}^{T} S_t^2}{\hat{\sigma}_\varepsilon^2}, \tag{5.18}$$

with $\hat{\sigma}_\varepsilon^2$ being an estimate of the error variance from step one. The authors
suggest using a Bartlett window $w(s, l) = 1 - s/(l+1)$ as an optimal weighting
function to estimate the long-run variance $\hat{\sigma}_\varepsilon^2$; that is,

$$\hat{\sigma}_\varepsilon^2 = s^2(l) = T^{-1} \sum_{t=1}^{T} \hat{\varepsilon}_t^2 + 2T^{-1} \sum_{s=1}^{l} 1 - \frac{s}{l+1} \sum_{t=s+1}^{T} \hat{\varepsilon}_t \hat{\varepsilon}_{t-1}. \tag{5.19}$$

The upper tail critical values of the level- and trend-stationarity versions are
given in Kwiatkowski et al. [1992].

The two test types are implemented as function `ur.kpss()` in the contributed package **urca** and as function `kpss.test()` in the contributed package **tseries**.[3] The implementation as in the package **urca** will be applied to

---

[3] The former function has been ported into the contributed package **fUnitRoots** as
`urkpssTest()`.

Table 5.15. KPSS tests for interest rates and nominal wages of the United States

Test Statistic	1%	5%	10%	
$\hat{\eta}_\mu$	0.13	0.35	0.46	0.74
$\hat{\eta}_\tau$	0.10	0.12	0.15	0.22

the Nelson and Plosser [1982] data set and thereby replicates some results in Kwiatkowski et al. [1992]. Besides the test type mu or tau for level-stationarity or trend-stationarity, the user has the option to set the Bartlett window parameter l via the argument lags either to short, which corresponds to the integer value $4 \times (T/100)^{1/4}$, to long, which is equivalent to the integer value $12 \times (T/100)^{1/4}$, or to nil. Alternatively, one can specify an integer value by providing the desired lag length via the argument use.lag. In R code 5.5, the level-stationary version is applied to the interest rate data, and the trend-stationary version of the test is applied to the logarithm of nominal wages. A lag length l of eight has been used by setting the functional argument use.lag accordingly.

R Code 5.5 KPSS test: Integration order for interest rates and nominal wages in the United States

```
library(urca)                                          1
data(nporg)                                            2
ir <- na.omit(nporg[, "bnd"])                          3
wg <- log(na.omit(nporg[, "wg.n"]))                    4
ir.kpss <- ur.kpss(ir, type = "mu", use.lag=8)         5
wg.kpss <- ur.kpss(wg, type = "tau", use.lag=8)        6
```

The null hypotheses of level- and trend-stationarity, respectively, cannot be rejected for both series, as shown in Table 5.15.

Summary

In this first chapter of Part II, various unit root tests have been applied to real data sets. The sequential testing strategy of the ADF test outlined in Section 3.2 has been applied to U.K. consumption. Because the data-generating process is unknown, it is recommended to go through these steps rather than merely apply one-test regressions as in Equations (5.1a)–(5.1c). Furthermore, a spherical error term should always be ensured by supplying sufficient lagged endogenous variables. Next, the Phillips-Perron test has been applied to the same data set. In principle, the difference between the two tests is that the latter uses a non-parametric correction that captures weak dependence and heterogeneity of the error process. As pointed out in Section 5.3, the relatively low power of both tests due to the fact that a unit root process is specified as the null hypothesis must be considered as a shortcoming. The ERS tests ameliorate this problem and should therefore be preferred. Furthermore, the nuisance parameters have different interpretations if either the null or the alternative hypothesis is true. The SP test addresses this problem explicitly and allows inclusion of higher polynomials in the deterministic part. However, all tests suffer from an ill-specified null hypothesis. The KPSS test, as a test for stationarity, correctly addresses the hypothesis specification from the viewpoint of conservative testing. Anyway, unfortunately there is no clear-cut answer to the question of which test should be applied to a data set. A combination of some of the above-mentioned tests with the inclusion of opposing null hypotheses therefore seems to be a pragmatic approach in practice.

Exercises

1. Determine the order of integration for the income and wealth series contained in the data set `Raotbl3` with the ADF and the PP tests.
2. Apply the ERS tests to the Nelson and Plosser data set contained in `nporg`, and compare your findings with the ones in Nelson and Plosser [1982].
3. Replicate the results in Kwiatkowski et al. [1992], and again compare them with the results in Nelson and Plosser [1982].
4. For response surface regression for the ERS tests P_T^μ and P_T^τ:
   (a) First, write a function that displays the critical values of the P_T^μ and P_T^τ statistics as provided in Table I of Elliott et al. [1996].
   (b) Next, write a function for conducting a Monte Carlo simulation of the P_T^μ and P_T^τ statistics for finite samples.
   (c) Fit a response surface regression to your results from Exercise 4(b).
   (d) Finally, compare the critical values implied from the response surface regression with the ones provided in Table I of Elliott et al. [1996] for selected significance levels and sample sizes.
5. Complete the following table:

Data Set	ADF	PP	ERS	SP	KPSS
(Raotbl3) lc	$I(1)$	$I(1)$			
(nporg) gnp.r			$I(1)$		
(nporg) gnp.n				$I(1)$	
(nporg) bnd					$I(0)$
(nporg) wg.n					$I(0)$

6

Further Considerations

In Chapter 5, various unit root tests were introduced and compared with each other. This chapter deals with two further topics. First, the case of structural breaks in a time series and how this affects the inference about the degree of integratedness is considered. Second, the issue of seasonal unit roots is discussed, as it was only briefly touched on in Section 3.2.

6.1 Stable Autoregressive Processes with Structural Breaks

Recall from Section 3.2 the random walk with drift model as in Equation (3.7). It has been argued that the drift parameter μ can be viewed as a deterministic trend given the final form

$$y_t = \mu t + y_0 + \sum_{i=1}^{t} \varepsilon_t. \tag{6.1}$$

Now suppose that the series is contaminated by a *structural break*. Such an occurrence can be caused by new legislation that affects the economy or by a redefinition of the data series; *e.g.*, a new definition for counting the unemployed has been decreed. One can distinguish two different ways for how such a structural shift impacts a series. Either the break occurs at only one point in time and then lasts for the remaining periods of the sample or it influences the series only in one particular period. In practice, such structural shifts are modeled by introducing dummy variables. In the former case, the structural shift is modeled as a step dummy variable that is zero before the break date and unity afterward. The latter is referred to as a pulse intervention, and the dummy variable is only unity at the break date and zero otherwise. Either way, if the series is $I(1)$, such a structural shift will have a lasting effect on the series. Consider the data-generating process

$$y_t = \mu + \delta D_t + y_{t-1} + \varepsilon_t, \tag{6.2}$$

where D_t assigns a pulse dummy variable that is defined as

$$D_t = \begin{cases} 1 & t = \tau, \\ 0 & \text{otherwise}, \end{cases}$$

where τ assigns the break date. Even though the break occurs in only one period, it will have a lasting effect on the series, as can be seen by calculating the final form of Equation (6.2), which is given by

$$y_t = \mu + \delta S_t + y_0 + \sum_{i=1}^{t} \varepsilon_t, \qquad (6.3)$$

and S_t is

$$S_t = \begin{cases} 1 & t \geq \tau, \\ 0 & \text{otherwise.} \end{cases}$$

In R code 6.1, two random walks with drift have been generated from the same sequence of disturbances with size 500. The second process has been affected at observation 250 with a pulse dummy, defined as object S (see command lines 5 and 9). The two series are plotted in Figure 6.1.

R Code 6.1 Random walk with drift and structural break

```
set.seed(123456)                                      1
e <- rnorm(500)                                       2
## trend                                              3
trd <- 1:500                                          4
S <- c(rep(0, 249), rep(1, 251))                      5
## random walk with drift                             6
y1 <- 0.1*trd + cumsum(e)                             7
## random walk with drift and shift                   8
y2 <- 0.1*trd + 10*S + cumsum(e)                      9
```

The difficulty in statistically distinguishing an $I(1)$-series from a stable $I(0)$ one develops if the latter is contaminated by a structural shift. Hence, the inference drawn from a Dickey-Fuller-type test becomes unreliable in the case of a potential structural break. This has been shown by Perron [1989], Perron [1990], Perron [1993], and Perron and Vogelsang [1992]. In Perron [1989], three different kinds of models are considered, where the structural break point is assumed to be known:

$$\text{Model (A):} \quad y_t = \mu + dD(T_\tau) + y_{t-1} + \varepsilon_t, \qquad (6.4a)$$
$$\text{Model (B):} \quad y_t = \mu_1 + (\mu_2 - \mu_1)DU_t + y_{t-1} + \varepsilon_t, \qquad (6.4b)$$
$$\text{Model (C):} \quad y_t = \mu_1 + dD(T_\tau) + (\mu_2 - \mu_1)DU_t + y_{t-1} + \varepsilon_t, \qquad (6.4c)$$

where $1 < T_\tau < T$ assigns the a priori known break point, $D(T_\tau) = 1$ if $t = T_\tau + 1$ and 0 otherwise, and $DU_t = 1$ for $t > T_\tau$ and 0 otherwise. It is further assumed that the error process can be represented as $\phi(L)\varepsilon_t = \theta(L)\xi_t$

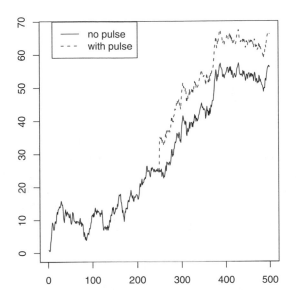

Fig. 6.1. Time series plot of random walk with drift and structural break

with ξ i.i.d., where $\phi(L)$ and $\theta(L)$ assign lag polynomials. In model (A), a one-time shift in the levels of the series is taken into account, whereas in model (B) a change in the rate of growth is allowed and model (C) is a combination of both. The model specifications for the trend-stationary alternative are

Model (A): $y_t = \mu_1 + \beta t + (\mu_2 - \mu_1)DU_t + \varepsilon_t,$ (6.5a)

Model (B): $y_t = \mu + \beta_1 t + (\beta_2 - \beta_1)DT_t^* + \varepsilon_t,$ (6.5b)

Model (C): $y_t = \mu + \beta_1 t + (\mu_2 - \mu_1)DU_t + (\beta_2 - \beta_1)DT_t^* + \varepsilon_t,$ (6.5c)

where $DT_t^* = t - T_\tau$ for $t > T_\tau$ and 0 otherwise.

The author proposed an adjusted, augmented Dickey-Fuller (ADF)-type test for the three models that is based on the following test regressions:

$$y_t = \hat{\mu}^A + \hat{\theta}^A DU_t + \hat{\beta}^A t + \hat{d}^A D(T_\tau)_t + \hat{\alpha}^A y_{t-1} + \sum_{i=1}^{k} \hat{c}_i^A \Delta y_{t-i} + \hat{\varepsilon}_t, \quad (6.6a)$$

$$y_t = \hat{\mu}^B + \hat{\beta}^B t + \hat{\gamma}^B DT_t^* + \hat{\alpha}^B y_{t-1} + \sum_{i=1}^{k} \hat{c}_i^B \Delta y_{t-i} + \hat{\varepsilon}_t, \quad (6.6b)$$

$$y_t = \hat{\mu}^C + \hat{\theta}^C DU_t + \hat{\beta}^C t + \hat{\gamma}^C DT_t^* + \hat{d}^C D(T_\tau)_t + \hat{\alpha}^C y_{t-1}$$
$$+ \sum_{i=1}^{k} \hat{c}_i^C \Delta y_{t-i} + \hat{\varepsilon}_t. \quad (6.6c)$$

The test statistic is the Student t ratio $t_{\hat{\alpha}^i}(\lambda)$ for $i = A, B, C$. Please note that this test statistic is now dependent on the fraction of the structural break point with respect to the total sample; *i.e.*, $\lambda = \frac{T_\tau}{T}$. The critical values of this test statistic are provided in Perron [1989] and/or Perron [1993]. The author applied these models to the data series used in Nelson and Plosser [1982] and concluded that most of the series no longer contains a unit root if the presence of a structural break is taken into account.

Zivot and Andrews [1992] pointed out that the risk of data mining exists if the break point is set exogenously by the researcher. They propose a test that circumvents this possibility by endogenously determining the most likely occurrence of a structural shift. By reanalyzing the data set used in Perron [1989], they found less evidence for rejecting the assumption of a unit root process. The estimation procedure they proposed is to choose the date of the structural shift for that point in time that gives the least favorable result for the null hypothesis of a random walk with drift. The test statistic is as in Perron [1989] the Student t ratio

$$t_{\hat{\alpha}^i}[\hat{\lambda}_{\text{inf}}^i] = \inf_{\lambda \in \Delta} t_{\hat{\alpha}^i}(\lambda) \quad \text{for} \quad i = A, B, C, \quad (6.7)$$

where Δ is a closed subset of $(0,1)$. Depending on which model is selected, the test statistic is inferred from one of the test regressions

$$y_t = \hat{\mu}^A + \hat{\theta}^A DU_t(\hat{\lambda}) + \hat{\beta}^A t + \hat{\alpha}^A y_{t-1} + \sum_{i=1}^{k} \hat{c}_i^A \Delta y_{t-i} + \hat{\varepsilon}_t, \quad (6.8a)$$

$$y_t = \hat{\mu}^B + \hat{\beta}^B t + \hat{\gamma}^B DT_t^*(\hat{\lambda}) + \hat{\alpha}^B y_{t-1} + \sum_{i=1}^{k} \hat{c}_i^B \Delta y_{t-i} + \hat{\varepsilon}_t, \quad (6.8b)$$

$$y_t = \hat{\mu}^C + \hat{\theta}^C DU_t(\hat{\lambda}) + \hat{\beta}^C t + \hat{\gamma}^C DT_t^*(\hat{\lambda}) + \hat{\alpha}^C y_{t-1}$$
$$+ \sum_{i=1}^{k} \hat{c}_i^C \Delta y_{t-i} + \hat{\varepsilon}_t, \quad (6.8c)$$

where $DU_t(\lambda) = 1$ if $t > T\lambda$ and 0 otherwise, and $DT_t^*(\lambda) = t - T\lambda$ for $t > T\lambda$ and 0 otherwise. Because now λ is estimated, one can no longer use the critical

Table 6.1. Zivot-Andrews: Test regression for nominal wages

Variable	Estimate	Std. Error	t-value	$\Pr(>|t|)$
(Intercept)	1.9878	0.3724	5.3375	0.0000
y.l1	0.6600	0.0641	10.2950	0.0000
trend	0.0173	0.0033	5.3190	0.0000
y.dl1	0.4979	0.1121	4.4411	0.0000
y.dl2	0.0557	0.1308	0.4262	0.6717
y.dl3	0.1494	0.1278	1.1691	0.2477
y.dl4	0.0611	0.1266	0.4826	0.6314
y.dl5	0.0061	0.1264	0.0484	0.9616
y.dl6	0.1419	0.1249	1.1364	0.2610
y.dl7	0.2671	0.1195	2.2358	0.0297
du	−0.1608	0.0387	−4.1577	0.0001

values as in Perron [1989] or Perron [1993], but the values published in Zivot and Andrews [1992] have to be used instead.

The Zivot and Andrews test is implemented as function ur.za() in the contributed package **urca**. In R code 6.2, this test is applied to the nominal and real wage series of the Nelson and Plosser data set. The test is applied to the natural logarithm of the two series. With the functional argument model, the type can be specified, in which intercept stands for model specification (A), trend corresponds to model type (B), and both is model type (C). The integer value of lag determines the number of lagged endogenous variables to be included in the test regression.

R Code 6.2 Unit roots and structural break: Zivot-Andrews test

```
library(urca)                                                      1
data(nporg)                                                        2
wg.n <- log(na.omit(nporg[, "wg.n"]))                              3
za.wg.n <- ur.za(wg.n, model = "intercept", lag = 7)              4
## plot(za.wg.n)                                                    5
wg.r <- log(na.omit(nporg[, "wg.r"]))                              6
za.wg.r <- ur.za(wg.r, model = "both", lag = 8)                    7
## plot(za.wg.r)                                                    8
```

The regression output (*i.e.*, the contents of the slot testreg) is displayed in Tables 6.1 and 6.2, respectively. The results of the test statistic are provided in Table 6.3. The unit root hypothesis must be rejected for the nominal wage series, given a significance level of 5%, whereas the unit root hypothesis cannot be rejected for the real wage series. The structural shift for the nominal wage

Table 6.2. Zivot-Andrews: Test regression for real wages

Variable	Estimate	Std. Error	t-value	Pr($>|t|$)
(Intercept)	2.5671	0.5327	4.8191	0.0000
y.l1	0.1146	0.1866	0.6141	0.5420
trend	0.0124	0.0028	4.4875	0.0000
y.dl1	0.6111	0.1662	3.6759	0.0006
y.dl2	0.3516	0.1686	2.0852	0.0423
y.dl3	0.4413	0.1568	2.8151	0.0070
y.dl4	0.2564	0.1453	1.7648	0.0838
y.dl5	0.1381	0.1346	1.0258	0.3100
y.dl6	0.0591	0.1262	0.4683	0.6416
y.dl7	0.1673	0.1201	1.3937	0.1697
y.dl8	0.1486	0.1210	1.2277	0.2254
du	0.0849	0.0196	4.3285	0.0001
dt	0.0081	0.0022	3.6819	0.0006

Table 6.3. Zivot-Andrews: Test statistics for real and nominal wages

Variable	Test Statistic	1%	5%	10%
wages, nominal		−5.30	−5.34	−4.80 −4.58
wages, real		−4.74	−5.57	−5.08 −4.82

series most likely occurred in period 30, which corresponds to the year 1929. The estimated break point is stored in the slot `bpoint`.

Finally, besides a `summary` method for objects of class `ur.za`, a `plot` method exists that depicts the path of the test statistic. The significance levels are drawn as separate lines, and in the case of a structural break, the break point is highlighted by a dashed vertical line. The graphs are displayed in Figures 6.2 and 6.3, respectively.

6.2 Seasonal Unit Roots

In Section 3.2, the topic of seasonal unit roots was briefly discussed. We will now investigate the issue of seasonal integration more thoroughly. This need originates because applied economists often need to construct models for seasonally unadjusted data. The reason for this is twofold. First, some data might be obviously seasonal in nature, and second, sometimes the utilization of seasonally adjusted data might distort the dynamics of an estimated model, as has been pointed out by Wallis [1974].

Recall the seasonal difference operator and its factorization, as shown in Equations (3.10a) and (3.10b). For quarterly data, this factorization would yield

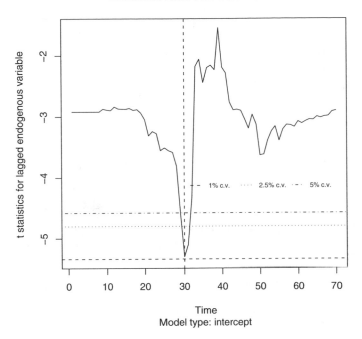

Fig. 6.2. Zivot-Andrews test statistic for nominal wages

$$(1 - L^4) = (1 - L)(1 + L)(1 - iL)(1 + iL), \tag{6.9}$$

where $\pm i$ are complex roots.[1] A seasonal quarterly process therefore has four possible roots, namely 1, -1, and $\pm i$. These roots correspond to different cycles in the time domain. The root 1 has a single-period cycle and is the zero-frequency root. The root -1 has a two-period cycle that implies for quarterly data a biannual cycle. Finally, the complex roots have a cycle of four periods that is equivalent to one cycle per year in quarterly data. The problem caused by the complex roots for quarterly data is that their effects are indistinguishable from each other. In Table 6.4, the cycles are summarized by the roots of the seasonal difference operator.

As mentioned in Section 3.2, the first attempts to test for seasonal unit roots, and probably the simplest, were suggested by Hasza and Fuller [1982] and Dickey et al. [1984]. The latter authors suggested the following test regression:

$$\Delta_s z_t = \delta_0 z_{t-1} + \sum_{i=1}^{k} \delta_i \Delta_s y_{t-i} + \varepsilon_t. \tag{6.10}$$

---

[1] For brevity, we consider quarterly data only. The factorizations for the other seasonal frequencies are provided in Franses and Hobijn [1997], for example.

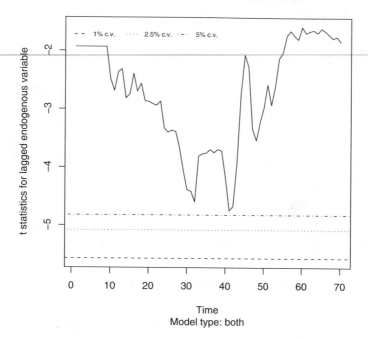

Fig. 6.3. Zivot-Andrews test statistic for real wages

Table 6.4. Cycles implied by the roots of the seasonal difference operator

Root $+1$	Root -1	Root $+i$	Root $-i$
Factor $(1-L)$	Factor $(1+L)$	Factor $(1-iL)$	Factor $(1+iL)$
$y_{t+1} = y_t$	$y_{t+1} = -y_t$	$y_{t+1} = iy_t$	$y_{t+1} = -iy_t$
	$y_{t+2} = -(y_{t+1}) = y_t$	$y_{t+2} = i(y_{t+1}) = -y_t$	$y_{t+2} = -i(y_{t+1}) = -y_t$
		$y_{t+3} = i(y_{t+2}) = -iy_t$	$y_{t+3} = -i(y_{t+2}) = iy_t$
		$y_{t+4} = i(y_{t+3}) = y_t$	$y_{t+4} = -i(y_{t+3}) = y_t$

The variable z_t is constructed by estimating the auxiliary regression

$$\Delta_s y_t = \sum_{i=1}^{h} \lambda_i \Delta_s y_{t-i} + \varepsilon_t, \tag{6.11}$$

which yields the coefficient estimates $\hat{\lambda}_1, \ldots, \hat{\lambda}_h$. The variable z_t is then constructed as

$$z_t = y_t - \sum_{i=1}^{h} \hat{\lambda}_i y_{t-i}. \tag{6.12}$$

The test for seasonal integration is then based on the Student t ratio for the ordinary least-squares estimate of the coefficient δ_0 in Equation (6.10).

Osborn et al. [1988] suggested replacing $\Delta_s z_t$ with $\Delta_s y_t$ as the dependent variable in Equation (6.10). Incidentally, if $h = 0$, this is equivalent with an ADF regression for the seasonal differences; $i.e.$,

$$\Delta_s y_t = \delta_0 y_{t-s} + \sum_{i=1}^{k} \delta_i \Delta_s y_{t-i} + \varepsilon_t. \tag{6.13}$$

The lag orders k and h should be determined similarly to the procedures proposed for the ADF test in Section 3.2. Furthermore, it should be noted that deterministic seasonal dummy variables can also be included in the test regression. The relevant critical values are provided in Osborn et al. [1988] and are dependent on the inclusion of such deterministic dummy variables and whether the data have been demeaned at the seasonal frequency. If the null hypothesis of the existence of a seasonal unit root is rejected for a large enough absolute t ratio, then one might conclude that stochastic seasonality is not present or that stochastic seasonality, which can be removed by using s-differences, does not exist. On the other hand, if the null hypothesis cannot be rejected, it is common practice to consider the order of non-seasonal differencing required to achieve stationarity instead of considering higher orders of seasonal differencing. Hence, one might consider as a data-generating process $SI(0,0)$ or $SI(0,1)$, or $SI(d,1)$ at most. To discriminate between $SI(0,1)$ and $SI(1,1)$, where the former is the new null hypothesis, one can estimate the ADF-type test regression

$$\Delta \Delta_s y_t = \delta_0 \Delta_s y_{t-1} + \sum_{i=1}^{k} \delta_i \Delta \Delta_s y_{t-i} + \varepsilon_t \tag{6.14}$$

and consider the t ratio for the hypothesis $\delta_0 = 0$. If this test statistic is insignificant, one takes into account $SI(2,1)$ as the new null hypothesis by estimating

$$\Delta \Delta \Delta_s y_t = \delta_0 \Delta \Delta_s y_{t-1} + \sum_{i=1}^{k} \delta_i \Delta \Delta \Delta_s y_{t-i} + \varepsilon_t \tag{6.15}$$

and so forth.

The deficiency of the test regression proposed by Osborn et al. [1988] is that it does not test all possible unit roots in a seasonal process (see Table 6.4). Hylleberg et al. [1990] suggested a test that allows for cyclical movements at different frequencies and takes the factorization of the seasonal lag polynomial as in Table 6.4 explicitly into account (the HEGY test). For quarterly data, they propose the test regression

$$\Delta_4 y_t = \sum_{i=1}^{4} \pi_i Y_{i,t-i} + \varepsilon_t, \tag{6.16}$$

where the regressors $Y_{i,t}$ for $i = 1, \ldots, 4$ are constructed as

$$Y_{1,t} = (1 + L)(1 + L^2)y_t = y_t + y_{t-1} + y_{t-2} + y_{t-3}, \qquad (6.17a)$$

$$Y_{2,t} = -(1 - L)(1 + L^2)y_t = -y_t + y_{t-1} - y_{t-2} + y_{t+3}, \qquad (6.17b)$$

$$Y_{3,t} = -(1 - L)(1 + L)y_t = -y_t + y_{t+2}, \qquad (6.17c)$$

$$Y_{4,t} = -(L)(1 - L)(1 + L)y_t = Y_{3,t-1} = -y_{t-1} + y_{t-3}. \qquad (6.17d)$$

The null hypothesis of seasonal integration implies that the coefficients π_i for $i = 1, \ldots, 4$ are equal to zero. As outlined, each π_i has a different interpretation. If only π_1 is significantly negative, then there is no non-seasonal stochastic stationary component in the data-generating process. Likewise, if only π_2 is significant, then there is no evidence of a biannual cycle in the data. Finally, the significance of π_3 and π_4 can be tested jointly with a Lagrange-multiplier F test. To put it differently, the existence of unit roots at the zero, biannual, and annual frequencies correspond to $\pi_1 = 0$, $\pi_2 = 0$, and $\pi_3 = \pi_4 = 0$, respectively. It should be noted that deterministic terms, such as an intercept, a trend, seasonal dummy variables, or a combination of these, as well as lagged seasonal differences, can be added to the test regression (6.16). That is, the general test regression is given by

$$\Delta_4 y_t = \pi_0 + \sum_{i=1}^{4} \pi_i Y_{i,t-1} + \sum_{i=1}^{3} \beta_i DS_{i,t} + \sum_{i=1}^{k} \delta_i \Delta_4 y_{t-1} + \gamma t + \varepsilon_t, \qquad (6.18)$$

where $DS_{i,t}$ assign the seasonal dummy variables, π_0 the constant term, and t a time trend. The critical values are provided in Hylleberg et al. [1990] and are dependent on the specification chosen and the sample size.

The HEGY test is implemented as function `HEGY.test()` in the contributed package **uroot**.[2] The specification of the test regression is determined by the functional argument `itsd`, which is a three-element vector. If the first element of this vector is set to one, a constant is included. A zero as the first element of `itsd` refers to a regression without an intercept. Analogously, if the second element of `itsd` is set to one, a linear trend is included, and a zero indicates its omission. The inclusion of seasonal dummy variables is controlled by the third element of `itsd`, which itself is a vector containing the dummy variables to be included in the test regression. The inclusion of lagged seasonal differences is set by the argument `selectlags`, which can be a specific order or an automatic selection and is done according to either the

---

[2] Incidentally, the package is shipped with a graphical user interface that is launched by executing `urootgui()` from the console. Besides the HEGY function, it should be noted at this point that the ADF test with the option to include deterministic seasonal dummy variables is available as function `ADF.test()` as well as the tests proposed by Canova and Hansen [1995] as functions `CH.test()` and `CH.rectest()`. Other features of the package are the generation of a LATEX table containing the test results and a panel function for graphical inspection of the time series characteristics.

Akaike information criterion (AIC), Bayesian information criterion (BIC), or a Ljung-Box test, or only the significant lags are retained. Finally, additional regressors can be included by the argument `regvar`. This argument enables the researcher to model explicitly structural breaks in the seasonal means and increasing seasonal variation, as was suggested as an amendment to the HEGY test by Franses and Hobijn [1997]. The function returns an object of formal class `hegystat-class`. The test statistics are contained in the slots `hegycoefs` and `stats`, respectively.

In R code 6.3, the HEGY test is applied to the logarithm of real disposable income in the United Kingdom from 1955:Q1 until 1984:Q4. This series is contained in the data set `UKconinc` in the contributed package **urca** and was used in Hylleberg et al. [1990]. The authors have chosen the lags 1, 4, 5 in the augmented test regression (6.18) and have run a combination of the deterministic regressors (see command lines 5 to 9). The t ratios of π_i for $i = 1, \ldots, 4$ can be retrieved by `object@hegycoefs`, and the F statistics of the Lagrange-multiplier test are stored in `object@stats`. Finally, the significance of the deterministic regressors can be checked by inspecting `object@lmhegy`.

R Code 6.3 HEGY test for seasonal unit roots

```
library(urca)                                                    1
library(uroot)                                                   2
data(UKconinc)                                                   3
incl <- ts(UKconinc$incl, start = c(1955,1),                     4
           end = c(1984,4), frequency = 4)                       5
HEGY000 <- HEGY.test(wts = incl, itsd = c(0, 0, c(0)),           6
                     selectlags = list(mode = c(1,4,5)))         7
HEGY100 <- HEGY.test(wts = incl, itsd = c(1, 0, c(0)),           8
                     selectlags = list(mode = c(1,4,5)))         9
HEGY110 <- HEGY.test(wts = incl, itsd = c(1, 1, c(0)),          10
                     selectlags = list(mode = c(1,4,5)))        11
HEGY101 <- HEGY.test(wts = incl,                               12
                     itsd = c(1, 0, c(1, 2, 3)),               13
                     selectlags = list(mode = c(1,4,5)))        14
HEGY111 <- HEGY.test(wts = incl,                               15
                     itsd = c(1, 1, c(1, 2, 3)),               16
                     selectlags = list(mode = c(1,4,5)))        17
```

The test results are provided in Table 6.5, where in the first column the specification is given as I for the inclusion of an intercept, SD for the inclusion of seasonal dummies, and Tr abbreviates a linear trend. The reported t ratios for π_3 and π_4 as well as the F statistic $\pi_3 \cap \pi_4$ are all significant at the 5% level. Hence, the authors conclude that the null hypothesis of a unit root at the zero frequency cannot be rejected. However, the null hypothesis for the conjugate complex roots must be rejected.

Table 6.5. HEGY test: Real disposable income in the United Kingdom

Regression	$t{:}\pi_1$	$t{:}\pi_2$	$t{:}\pi_3$	$t{:}\pi_4$	$F{:}\pi_3 \cap \pi_4$
None	2.61	-1.44	-2.35	-2.51	5.68
I	-1.50	-1.46	-2.38	-2.51	5.75
I, SD	-1.56	-2.38	-4.19	-3.89	14.73
I, Tr	-2.73	-1.46	-2.52	-2.24	5.46
I, SD, TR	-2.48	-2.30	-4.28	-3.46	13.74

Summary

In this chapter, the analysis of integrated time series has been amended in two important ways. First, it has been shown that in the case of a structural break the test conclusion about the presence of a unit root in a time series can be biased toward accepting it. Therefore, if *a priori* knowledge of a structural shift exists or a break in the series is evident by eye-spotting, one should either use the Perron or the Zivot and Andrews test, respectively. Second, if the correlogram gives hindsight of seasonality in the time series, one should apply a seasonal unit root test. A complete analysis of a possibly integrated time series would therefore begin by testing whether breaks and/or stochastic seasonality exist, and depending on this outcome, unit root tests should be applied as shown in Chapter 5. After all, when the null hypothesis of a unit root must be rejected, it should be checked whether long-memory behavior is present as shown in Section 3.3.

Exercises

1. Write a function that displays the critical values for models of types (A), (B), and (C) as in Perron [1989].
2. Write a function that estimates the models of types (A), (B), and (C) as in Equations (6.5a)–(6.5c).
3. Combine your functions from Exercises 1 and 2 so that now the functions return the relevant critical values for a prior specified significance level.
4. Write a function for the seasonal unit root test proposed by Osborn et al. [1988].
5. Apply this function to the log of real disposable income in the United Kingdom as contained in the data set `UKconinc` and compare this with the results reported in Table 6.5.

Part III

Cointegration

7

Single-Equation Methods

This is the first chapter of the third and last part of this book. The cointegration methodology is first presented for the case of single-equation models. The Engle-Granger two-step procedure is demonstrated by estimating a consumption function and its error-correction form for the United Kingdom as in Holden and Perman [1994]. In the Section 7.2, the method proposed by Phillips and Ouliaris [1990] is applied to the same data set. The application and inferences of a vector error-correction model are saved for Chapter 8.

7.1 Engle-Granger Two-Step Procedure

Recall from Section 4.2 that the first step of the Engle-Granger two-step procedure consists of estimating the long-run relationship as in Equation (4.4). Holden and Perman [1994] applied this procedure to the estimation of a consumption function for the United Kingdom. The integration order of the consumption series was already discussed in Section 5.1, and the determination of the integration order of the income and wealth series was given as Exercise 1 in Chapter 5. In the following discussion, we will treat all series as $I(1)$, although the result for the wealth series is ambiguous.[1] The authors regressed consumption on income and wealth for the sample period from 1967:Q2 until 1991:Q2. In R code 7.1, the data set `Raotbl3` is loaded and the series are transformed into time series objects (see command lines 3 to 9). The selection of the sample period is easily achieved by the function `window()` in command line 11. By slightly diverging from the analysis as in Holden and Perman, the long-run relationships for each of the series (*i.e.*, consumption, income, and wealth), entered separately as endogenous variables, are simply estimated by ordinary least-squares (OLS) (see command lines 13 to 15).

The residuals of these three long-run relationships are stored as objects `error.lc`, `error.li`, and `error.lw`, respectively. An augmented Dickey-Fuller (ADF)-type test is applied to the residuals of each equation for testing whether

---

[1] When a broken trend is allowed in the data-generating process for the wealth series, the authors concluded that the unit root hypothesis must be rejected on the basis of the test proposed by Perron [1989]. This finding is confirmed by applying the Zivot and Andrews [1992] test for a model with a constant, trend, and four lags (see Section 6.1 for a discussion of both tests).

the variables are cointegrated or not (see command lines 22 to 24). Please note that one must now use the critical values found in MacKinnon [1991] or Engle and Yoo [1987]. The test statistics imply cointegration for the consumption and income functions that are significant at the 5% level, given a critical value of -3.83, but not for the wealth equation, thereby stressing the finding that this series should be considered as stationary with a broken trend. Furthermore, the Jarque-Bera test indicates that the null hypothesis of normality cannot be rejected for all equations.

R Code 7.1 Engle-Granger: Long-run relationship of consumption, income, and wealth in the United Kingdom

```
library(tseries)                                              1
library(urca)                                                 2
data(Raotbl3)                                                 3
attach(Raotbl3)                                               4
lc <- ts(lc, start=c(1966,4), end=c(1991,2),                  5
         frequency=4)                                         6
li <- ts(li, start=c(1966,4), end=c(1991,2),                  7
         frequency=4)                                         8
lw <- ts(lw, start=c(1966,4), end=c(1991,2),                  9
         frequency=4)                                        10
ukcons <- window(cbind(lc, li, lw), start=c(1967, 2),        11
                 end=c(1991,2))                              12
lc.eq <- summary(lm(lc ~ li + lw, data=ukcons))             13
li.eq <- summary(lm(li ~ lc + lw, data=ukcons))             14
lw.eq <- summary(lm(lw ~ li + lc, data=ukcons))             15
error.lc <- ts(resid(lc.eq), start=c(1967,2),               16
               end=c(1991,2), frequency=4)                   17
error.li <- ts(resid(li.eq), start=c(1967,2),               18
               end=c(1991,2), frequency=4)                   19
error.lw <- ts(resid(lw.eq), start=c(1967,2),               20
               end=c(1991,2), frequency=4)                   21
ci.lc <- ur.df(error.lc, lags=1, type='none')               22
ci.li <- ur.df(error.li, lags=1, type='none')               23
ci.lw <- ur.df(error.lw, lags=1, type='none')               24
jb.lc <- jarque.bera.test(error.lc)                         25
jb.li <- jarque.bera.test(error.li)                         26
jb.lw <- jarque.bera.test(error.lw)                         27
```

In the next step, the error-correction models (ECMs) for the consumption and income functions are specified as in Equations (4.5a) and (4.5b). In R code 7.2, the necessary first differences of the series and its lagged values are created, as well as the series for the error term lagged by one period.

Table 7.1. Engle-Granger: Cointegration test

Variable	ADF	JB	p-value
Consumption	−4.14	0.66	0.72
Income	−4.06	0.07	0.97
Wealth	−2.71	3.25	0.20

R **Code 7.2** Engle-Granger: ECM for consumption and income of the United Kingdom

```
ukcons2 <- ts(embed(diff(ukcons), dim=2),           1
          start=c(1967,4), freq=4)                    2
colnames(ukcons2) <- c('lc.d', 'li.d', 'lw.d',        3
          'lc.d1', 'li.d1', 'lw.d1')                  4
error.ecm1 <- window(lag(error.lc, k=-1),             5
          start=c(1967,4), end=c(1991, 2))            6
error.ecm2 <- window(lag(error.li, k=-1),             7
          start=c(1967,4), end=c(1991, 2))            8
ecm.eq1 <- lm(lc.d ~ error.ecm1 + lc.d1 + li.d1 + lw.d1,   9
          data=ukcons2)                               10
ecm.eq2 <- lm(li.d ~ error.ecm2 + lc.d1 + li.d1 + lw.d1,   11
          data=ukcons2)                               12
```

The regression results for both ECMs are depicted in Tables 7.2 and 7.3. It should be restressed at this point that if two series are cointegrated, then there should be Granger-causation in at least one direction. That is, at least one coefficient of the error term should enter Equations (4.5a) or (4.5b) significantly and with the correct sign (*i.e.*, negative). Hence, even if the lagged differences of the income and consumption regressors do not enter significantly, the levels might through the residuals and hence Granger-cause consumption and/or income. The coefficient of the error-correction term in the consumption function does not enter significantly and has the wrong sign (see Table 7.2). On the contrary, the error-correction term does enter significantly and has the correct sign in the income equation (see Table 7.3). The error of the last period is worked off by one half, although the lagged differences of the remaining regressors do not enter significantly into the ECM. These results imply Granger-causation from consumption to income.

7.2 Phillips-Ouliaris Method

In Section 7.1 and Section 4.2, it has been shown that the second step of the Engle-Granger method is an ADF-type test applied to the residuals of the

Table 7.2. Engle-Granger: ECM for the consumption function

Variable	Estimate	Std. Error	t-value	Pr($>|t|$)
(Intercept)	0.0058	0.0015	3.8556	0.0002
error.ecm1	0.0625	0.0984	0.6354	0.5268
lc.d1	−0.2856	0.1158	−2.4655	0.0156
li.d1	0.2614	0.0864	3.0270	0.0032
lw.d1	0.0827	0.0317	2.6097	0.0106

Table 7.3. Engle-Granger: ECM for the income function

Variable	Estimate	Std. Error	t-value	Pr($>|t|$)
(Intercept)	0.0066	0.0019	3.5346	0.0006
error.ecm2	−0.5395	0.1142	−4.7236	0.0000
lc.d1	−0.1496	0.1464	−1.0218	0.3096
li.d1	−0.0060	0.1085	−0.0556	0.9558
lw.d1	0.0627	0.0398	1.5753	0.1187

long-run equation. Phillips and Ouliaris [1990] introduced two residual-based tests, namely a variance ratio and a multivariate trace statistic. The latter of these tests has the advantage that it is invariant to normalization (*i.e.*, which variable is taken as endogenous). Both tests are based on the residuals of the first-order vector autoregression

$$z_t = \hat{\Pi} z_{t-1} + \hat{\xi} i_t, \tag{7.1}$$

where z_t is partitioned as $z_t = (y_t, x_t')$ with a dimension of x_t equal to $m = n + 1$. The variance ratio statistic $\hat{P}_u$ is then defined as

$$\hat{P}_u = \frac{T\hat{\omega}_{11\cdot 2}}{T^{-1}\sum_{t=1}^{T} \hat{u}_t^2}, \tag{7.2}$$

where $\hat{u}_t$ are the residuals of the long-run equation $y_t = \hat{\beta}' x_t + \hat{u}_t$. The conditional covariance $\hat{\omega}_{11\cdot 2}$ is derived from the covariance matrix $\hat{\Omega}$ of $\hat{\xi}_t$ (*i.e.*, the residuals of Equation (7.1)) and is defined as

$$\hat{\omega}_{11\cdot 2} = \hat{\omega}_{11} - \hat{\omega}_{21}' \hat{\Omega}_{22}^{-1} \hat{\omega}_{21}, \tag{7.3}$$

where the covariance matrix $\hat{\Omega}$ has been partitioned as

$$\hat{\Omega} = \begin{bmatrix} \hat{\omega}_{11} & \hat{\omega}_{21} \\ \hat{\omega}_{21} & \hat{\Omega}_{22} \end{bmatrix}, \tag{7.4}$$

and is estimated as

$$\hat{\Omega} = T^{-1} \sum_{t=1}^{T} \hat{\xi}_t' \hat{\xi}_t + T^{-1} \sum_{s=1}^{l} w_{sl} \sum_{t=1}^{T} (\hat{\xi}_t \hat{\xi}_{t-s}' + \hat{\xi}_{t-s} \hat{\xi}_t'), \qquad (7.5)$$

with weighting function $w_{sl} = 1 - s/(l+1)$. Therefore, the variance ratio statistic measures the size of the residual variance from the cointegrating regression of y_t on x_t against that of the conditional variance of y_t given x_t. In the case of cointegration, the test statistic should stabilize to a constant, whereas if a spurious relationship is present, this would be reflected in a divergent variance of the long-run equation residuals from the conditional variance. Critical values of this test statistic have been tabulated in Phillips and Ouliaris [1990].

The multivariate trace statistic, denoted as $\hat{P}_z$, is defined as

$$\hat{P}_z = T\,tr(\hat{\Omega} M_{zz}^{-1}), \qquad (7.6)$$

with $M_{zz} = t^{-1} \sum_{t=1}^{T} z_t z_t'$ and $\hat{\Omega}$ estimated as in Equation (7.5). Critical values for this test statistic are provided in Phillips and Ouliaris [1990], too. The null hypothesis is that no cointegration relationship exists.

Both tests are implemented in the function `ca.po()` in the contributed package **urca**. Besides the specification of the test type, the inclusion of deterministic regressors can be set via the argument `demean`, and the lag length for estimating the long-run variance-covariance matrix $\hat{\Omega}$ can be set with the argument `lag`. Because a matrix inversion is needed in the calculation of the test statistics, one can pass a tolerance level to the implicitly used function `solve()` via the argument `tol`. The default value is `NULL`.

In R code 7.3, both test types are applied to the same data set as before. The results are provided in Table 7.4. The variance ratio test statistic does not indicate a spurious relationship. This should be no surprise because the first column of the data set `ukcons` is the consumption series. Therefore, the test conclusion is the same as when using the two-step Engle-Granger procedure.

However, matters are different if one uses the $\hat{P}_z$ statistic. From the inclusion of the wealth series, which is considered as stationary around a broken trend line, the statistic indicates a spurious relationship at the 5% level. Please note that normalization of the long-run equation does not affect this test statistic.

Up to now we have only discussed single-equation methods and how such methods can be fairly easily applied in R. One deficiency of these methods is that one can only estimate a single cointegration relationship. However, if one deals with more than two time series, it is possible that more than only one cointegrating relationship exists, as has been pointed out in Section 4.3. The estimation and inference of VECMs are the subject of the next chapter.

Table 7.4. Phillips-Ouliaris: Cointegration test

Variable	Test Statistic	10%	5%	1%
$\hat{P}_u$	58.91	33.70	40.53	53.87
$\hat{P}_z$	88.03	80.20	89.76	109.45

R Code 7.3 Phillips-Ouliaris: Long-run relationship of consumption, income, and wealth in the United Kingdom

```
library(urca)                                                     1
data(Raotbl3)                                                     2
attach(Raotbl3)                                                   3
lc <- ts(lc, start=c(1966,4), end=c(1991,2),                     4
         frequency=4)                                             5
li <- ts(li, start=c(1966,4), end=c(1991,2),                     6
         frequency=4)                                             7
lw <- ts(lw, start=c(1966,4), end=c(1991,2),                     8
         frequency=4)                                             9
ukcons <- window(cbind(lc, li, lw), start=c(1967, 2),           10
              end=c(1991,2))                                     11
pu.test <- summary(ca.po(ukcons, demean='const',               12
              type='Pu'))                                        13
pz.test <- summary(ca.po(ukcons, demean='const',               14
              type='Pz'))                                        15
```

Summary

In this first chapter of Part III, two single-equation methods have been presented. The advantage of the Engle-Granger two-step procedure is its ease of implementation. However, the results are dependent on how the long-run equation is specified. In most cases, it might be obvious which variable enters on the left-hand side of the equation; *i.e.*, to which variable the cointegrating vector should be normalized. Unfortunately, this is only true in most cases, and, as anecdotal evidence, an income function rather than a consumption function could have been specified as an ECM in R code 7.2. It is therefore advisable to employ the cointegration test of Phillips and Ouliaris, which is irrelevant to normalization.

As mentioned, the insights gained with respect to the cointegrating relationship are limited in the case of more than two variables. The next chapter is therefore dedicated to the inference in cointegrated systems.

Exercises

1. Consider the data sets `Raotbl1` and `Raotbl2` in the contributed package **urca**. Your goal is to specify the ECM for real money demand functions by using different monetary aggregates.
   (a) Determine the integration order of the series first.
   (b) Estimate the long-run equations.
   (c) Can you find cointegration relations for the different money demand functions?
   (d) Specify the ECM and interpret your results with respect to the error-correction term.
2. Consider the data set `Raotbl6` in the contributed package **urca**. Specify a Phillips-curve model in error-correction form as in Mehra [1994].
   (a) Determine the integration order of the price level, unit labor cost, and output gap variable first.
   (b) Estimate the long-run equation and test for cointegration. Employ the Phillips-Ouliaris tests, too.
   (c) Specify an ECM, and discuss your findings.

8

Multiple-Equation Methods

In this chapter, the powerful tool of likelihood-based inference in cointegrated vector autoregressive models (VECMs) is discussed. In the first section, the specification and assumptions of a VECM are introduced. In the following sections, the problems of determining the cointegration rank, testing for weak exogeneity, and testing of various restrictions placed on the cointegrating vectors are discussed. The topic of VECMs that are contaminated by a one-time structural shift and how this kind of model can be estimated are presented. This chapter concludes with an exposition of structural vector error-correction models.

8.1 The Vector Error-Correction Model

8.1.1 Specification and Assumptions

In this section, the results in Johansen and Juselius [1992] are replicated. In this article, the authors test structural hypotheses in a multivariate cointegration context of the *purchasing power parity* and the *uncovered interest parity* for the United Kingdom. They use quarterly data spanning a range from 1972:Q1 to 1987:Q2. As variables, p_1, the U.K. wholesale price index; p_2, the trade-weighted foreign wholesale price index; e_{12}, the U.K. effective exchange rate; i_1, the three-month treasury bill rate in the United Kingdom; and i_2, the three-month Eurodollar interest rate, enter into the VECM. To cope with the oil crisis at the beginning of the sample period, the world oil price (contemporaneously and lagged once), denoted as $doilp_0$ and $doilp_1$, respectively, is included as an exogenous regressor, too. These variables, expressed in natural logarithms, are included in the data set UKpppuip contained in the package **urca** and are depicted in Figure 8.1.

As a preliminary model, Johansen and Juselius settled on the specification

$$y_t = \Gamma_1 \Delta y_{t-1} + c_0 \Delta x_t + c_1 \Delta x_{t-1} + \Pi y_{t-2} + \mu + \Phi D_t + \varepsilon_t, \qquad (8.1)$$

where the vector y_t contains as elements $(p_1, p_2, e_{12}, i_1, i_2)'$ and x_t assigns the model exogenous world oil price $doilp_0$. In the matrix D_t, centered seasonal dummy variables are included, and the vector μ is a vector of constants. The five-dimensional error process ε_t is assumed to be i.i.d. as $\mathcal{N}(0, \Sigma)$ for $t = 1, \ldots, T$. This specification is the long-run form of a VECM (see Equation (4.8)).

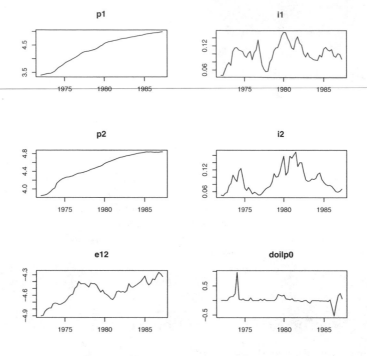

Fig. 8.1. Graphical display of purchasing power parity and uncovered interest rate parity for the United Kingdom

8.1.2 Determining the Cointegration Rank

Johansen and Juselius start by determining the cointegration rank. Because inferences on the cointegration space spanned by its vectors are dependent on whether or not linear trends exist in the data, they argued by ocular econometrics and logical reasoning that the price series have a linear trend that is consistent with the steady-state assumption of constant nominal price growth as implied by economic theory, and therefore the vector μ can be estimated without imposing any restrictions.[1]

---

[1] In the case of linear trends, the constant vector μ can be partitioned as $\mu = \alpha\beta_0 + \alpha_\perp\gamma$, where β_0 is an $r \times 1$ vector of intercepts in the cointegration relations, $\alpha_\perp$ is a $K \times (K-r)$ matrix of full rank perpendicular to the columns of α, and γ is a $(K-r) \times 1$ vector of linear trend slopes. Therefore, the alternative hypothesis is $\alpha_\perp\gamma = 0$ and can be tested as shown in Johansen and Juselius [1990]. This linear trend test is implemented in the package **urca** as function lttest().

In R code 8.1, the hypothesis $\mathcal{H}_1(r)$: $\boldsymbol{\Pi} = \boldsymbol{\alpha\beta'}$ (*i.e.*, that $\boldsymbol{\Pi}$ is of re-duced rank) is tested with the trace and the maximal eigenvalue statistic (see Equations (4.11) and (4.14)).[2]

R Code 8.1 Johansen-Juselius: Unrestricted cointegration

```
library(urca)                                               1
data(UKpppuip)                                              2
names(UKpppuip)                                             3
attach(UKpppuip)                                            4
dat1 <- cbind(p1, p2, e12, i1, i2)                          5
dat2 <- cbind(doilp0, doilp1)                               6
args('ca.jo')                                               7
H1 <- ca.jo(dat1, type = 'trace', K = 2, season = 4,        8
            dumvar = dat2)                                  9
H1.trace <- summary(ca.jo(dat1, type = 'trace', K = 2,      10
                    season = 4, dumvar = dat2))             11
H1.eigen <- summary(ca.jo(dat1, type = 'eigen', K = 2,      12
                    season = 4, dumvar = dat2))             13
```

Before the results of these tests are discussed, the arguments of the function `ca.jo()` should be presented briefly. The data set is provided by x, and the test type is either **eigen** or **trace** for the maximal eigenvalue statistic or the trace statistic, respectively, where the default is the former. Whether no deterministic term, a constant, or a trend should be included in the coin-tegration relations can be set by the argument **ecdet**. The decision as to whether the long-run or transitory form of the VECM should be estimated is determined by the argument **spec**. The default is **spec="longrun"**. The inclusion of centered seasonal dummy variables can be set by providing the corresponding seasonality as an integer; *e.g.*, **season** = 4 for quarterly data. Model exogenous regressors can be provided by setting **dumvar** accordingly.

In Tables 8.1 and 8.2, the results of the two tests are given. If considering the maximal eigenvalue statistic, the hypothesis of no cointegration cannot be rejected at the 5% level.[3] However, the trace statistic indicates a cointegration

---

[2] Incidentally, the internal examples of the functions for estimating and testing a VECM in the package **urca** are a replication of Johansen and Juselius [1990]; *i.e.*, the analysis of money demand functions for Denmark and Finland. For example, by typing **example(ca.jo())**, the results of determining the cointegration rank in this study are displayed. The reader is encouraged to work through these examples to foster understanding and comprehension of the method and the tools available. It is of course best accomplished by having a copy of the above-cited article at hand.

[3] The critical values in the article differ slightly from the ones that are returned by the **summary** method of **ca.jo()**. In function **ca.jo()**, the critical values provided

Table 8.1. Cointegration rank: Maximal eigenvalue statistic

Rank	Test Statistic	10%	5%	1%
$r <= 4$	5.19	6.50	8.18	11.65
$r <= 3$	6.48	12.91	14.90	19.19
$r <= 2$	17.59	18.90	21.07	25.75
$r <= 1$	20.16	24.78	27.14	32.14
$r = 0$	31.33	30.84	33.32	38.78

space of $r = 2$, given a 5% significance level. Hence, the two tests yield contradictary conclusions about the cointegration rank. The decision about the cointegration rank is complicated even more by the fact that the estimated second and third eigenvalues are approximately equal (0.407, 0.285, 0.254, 0.102, 0.083) and therefore suggest a third stationary linear combination. The eigenvalues are in the slot lambda of an object adhering to class ca.jo. To settle for a working assumption about the cointegration rank, Johansen and Juselius investigated the $\hat{\beta}$ and $\hat{\alpha}$ matrices as well as the estimated cointegration relations $\hat{\beta}'_i y_t$ and those that are corrected for short-term influences, $\hat{\beta}'_i R_{1t}$. The $\hat{\beta}$ and $\hat{\alpha}$ matrices are stored in the slots V and W, respectively, and the matrix R_1 in the slot RK of a class ca.jo object. For ease of comparison with Table 3 in Johansen and Juselius [1992], the elements in the cointegration and loadings matrix have been normalized accordingly and are displayed in Tables 8.3 and 8.4.

In R code 8.2, the commands for calculating the above-mentioned figures are displayed. The authors argued that the values of $\hat{\alpha}_{i.2}$ for $i = 1, 2, 3$ are close to zero for the second cointegration vector, and therefore the low estimated value of the second eigenvalue $\hat{\lambda}_2$ can be attributed to this fact. Furthermore, the power of the test is low in cases where the cointegration relation is close to the non-stationary boundary. This artifact can be the result of a slow speed of adjustment, as is often the case in empirical work because of transaction costs and other obstacles that place a hindrance on a quick equilibrium adjustment. Johansen and Juselius investigated the cointegration relationships visually. The first two of them are depicted in Figure 8.2. In the case of $r = 2$, the first two cointegration relationships should behave like stationary processes. However, because of short-run influences that overlay the adjustment process, this picture can be camouflaged. Hence, the authors also analyze the adjustment paths $\hat{\beta}'_i R_{1t}$ that take the short-run dynamics into account. Based on the test results, the elements in the $\hat{\alpha}$ matrix, and the shape of the cointegration relation paths, Johansen and Juselius decided to stick with a cointegration order of $r = 2$.

---

by Osterwald-Lenum [1992] are used. Osterwald-Lenum [1992] provides values for models of higher dimension and also for VECM specifications that allow the inclusion of a trend in the cointegration relations, for instance.

Table 8.2. Cointegration rank: Trace statistic ✓

Rank	Test Statistic	10%	5%	1%
$r <= 4$	5.19	6.50	8.18	11.65
$r <= 3$	11.67	15.66	17.95	23.52
$r <= 2$	29.26	28.71	31.52	37.22
$r <= 1$	49.42	45.23	48.28	55.43
$r = 0$	80.75	66.49	70.60	78.87

R **Code 8.2** $\mathcal{H}_1$ model: Transformations and cointegration relations

```
beta <- H1@V                                                    1
beta[,2] <- beta[,2]/beta[4,2]                                  2
beta[,3] <- beta[,3]/beta[4,3]                                  3
alpha <- H1@PI%*%solve(t(beta))                                 4
beta1 <- cbind(beta[,1:2], H1@V[,3:5])                          5
ci.1 <- ts((H1@x%*%beta1)[-c(1,2),], start=c(1972, 3),         6
        end=c(1987, 2), frequency=4)                            7
ci.2 <- ts(H1@RK%*%beta1, start=c(1972, 3),                    8
        end=c(1987, 2), frequency=4)                            9
```

Table 8.3. $\mathcal{H}_1$ model: Eigenvectors

Variable	$\hat{\beta}_{0.1}$	$\hat{\beta}_{0.2}$	$\hat{v}_3$	$\hat{v}_4$	$\hat{v}_5$
p1.l2	1.00	0.03	0.36	1.00	1.00
p2.l2	−0.91	−0.03	−0.46	−2.40	−1.45
e12.l2	−0.93	−0.10	0.41	1.12	−0.48
i1.l2	−3.37	1.00	1.00	−0.41	2.28
i2.l2	−1.89	−0.93	−1.03	2.99	0.76

Table 8.4. $\mathcal{H}_1$ model: Weights

Variable	$\hat{\alpha}_{0.1}$	$\hat{\alpha}_{0.2}$	$\hat{w}_3$	$\hat{w}_4$	$\hat{w}_5$
p1.d	−0.07	0.04	−0.01	0.00	−0.01
p2.d	−0.02	0.00	−0.04	0.01	0.01
e12.d	0.10	−0.01	−0.15	−0.04	−0.05
i1.d	0.03	−0.15	−0.03	0.01	−0.02
i2.d	0.06	0.29	0.01	0.03	−0.01

Finally, in Table 8.5, the estimated $\hat{\Pi}$ is displayed. This matrix measures the combined effects of the two cointegration relations. The purchasing power parity hypothesis postulates a relationship between the two price indices and

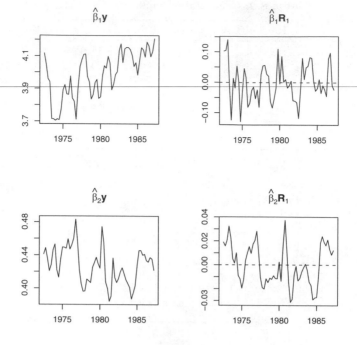

Fig. 8.2. Graphical display of the first two cointegration relations

Table 8.5. $\mathcal{H}_1(2)$ model: Coefficient matrix of the lagged variables in levels, $\hat{\Pi}$

Variable	p1.l2	p2.l2	e12.l2	i1.l2	i2.l2
p1.d	−0.067	0.061	0.060	0.272	0.090
p2.d	−0.018	0.016	0.016	0.064	0.030
e12.d	0.101	−0.091	−0.093	−0.345	−0.186
i1.d	0.030	−0.026	−0.018	−0.263	0.072
i2.d	0.066	−0.062	−0.082	0.097	−0.382

the exchange rate of $(\alpha_i, -\alpha_i, -\alpha_i)$. These relations are closely fulfilled for the first, third, and fourth equations.

8.1.3 Testing for Weak Exogenity

By now it has been assumed that all components of $\boldsymbol{y}_t$ are treated as endogenous variables. But sometimes we are interested in treating some of these components as exogenous, such that model simulations can be carried out by providing alternative paths for the exogenous assumed variables. Hence, a test is required for the full system if the hypothesis that some components of $\boldsymbol{y}_t$

are exogenous is valid. The test idea is to see whether zero restrictions on the relevant rows of $\boldsymbol{\alpha}$ hold. The restrictions are contained in the $(K \times m)$ matrix $\boldsymbol{A}$ such that $\mathcal{H}_4 : \boldsymbol{\alpha} = \boldsymbol{A\Psi}$, whereby the elements of the matrix $\boldsymbol{\Psi}$ contain the new unrestricted loadings. Johansen and Juselius [1990] showed that this test can be traced down to an eigenvalue problem like the one given in Equation (4.12). For calculating the new sets of residuals that enter the conditional concentrated likelihood function, it is convenient to introduce the $(K \times (K - m))$ matrix $\boldsymbol{B}$, which is orthogonal to $\boldsymbol{A}$ such that $\boldsymbol{B'A} = \boldsymbol{0}$ or equivalently $\boldsymbol{B'\alpha} = \boldsymbol{0}$. Johansen and Juselius [1990] and Johansen [1995] then showed that the sets of residuals that enter the concentrated likelihood function are defined as

$$\tilde{\boldsymbol{R}}_{at} = \boldsymbol{A'R}_{0t} - \hat{\boldsymbol{S}}_{ab}\hat{\boldsymbol{S}}_{bb}^{-1}\boldsymbol{B'R}_{0t}, \tag{8.2a}$$

$$\tilde{\boldsymbol{R}}_{1t} = \boldsymbol{R}_{1t} - \hat{\boldsymbol{S}}_{1b}\hat{\boldsymbol{S}}_{bb}^{-1}\boldsymbol{B'R}_{0t}, \tag{8.2b}$$

where $\boldsymbol{R}_{0t}$ and $\boldsymbol{R}_{1t}$ and the product moment matrices $\hat{\boldsymbol{S}}_{ij}$ with $i, j = 0, 1$ are given in Equation (4.10) and the product moment matrices $\hat{\boldsymbol{S}}_{ab}$, $\hat{\boldsymbol{S}}_{bb}$, and $\hat{\boldsymbol{S}}_{1b}$ are defined as $\hat{\boldsymbol{S}}_{ab} = \boldsymbol{A'}\hat{\boldsymbol{S}}_{00}\boldsymbol{B}$, $\hat{\boldsymbol{S}}_{bb} = \boldsymbol{B'}\hat{\boldsymbol{S}}_{00}\boldsymbol{B}$, and $\hat{\boldsymbol{S}}_{1b} = \hat{\boldsymbol{S}}_{01}\boldsymbol{B}$. The relevant product moment matrices that enter the eigenvalue equation are defined as

$$\hat{\boldsymbol{S}}_{ij.b} = \frac{1}{T}\sum_{t=1}^{T}\tilde{\boldsymbol{R}}_{it}\tilde{\boldsymbol{R}}_{jt}' \text{ with } i, j = a, 1. \tag{8.3}$$

The maximum-likelihood estimator for $\boldsymbol{\beta}$ under the hypothesis $\mathcal{H}_4 : \boldsymbol{\alpha} = \boldsymbol{A\Psi}$ is defined as the solution to the eigenvalue equation

$$|\boldsymbol{\lambda}\hat{\boldsymbol{S}}_{11.b} - \hat{\boldsymbol{S}}_{1a.b}\hat{\boldsymbol{S}}_{aa.b}^{-1}\hat{\boldsymbol{S}}_{a1.b}| = 0 \tag{8.4}$$

with eigenvalues $\tilde{\lambda}_{4.1} > \tilde{\lambda}_{4.2} > \ldots > \tilde{\lambda}_{4.m} > \tilde{\lambda}_{4.m+1} = \ldots = \tilde{\lambda}_{4.K} = 0$, and the corresponding eigenvectors $\tilde{\boldsymbol{V}}_4 = (\tilde{\boldsymbol{v}}_{4.1}, \ldots, \tilde{\boldsymbol{v}}_{4.K})$ are normalized such that $\tilde{\boldsymbol{V}}_4'\hat{\boldsymbol{S}}_{11.b}\tilde{\boldsymbol{V}}_4 = \boldsymbol{I}$. The weighting matrix $\hat{\boldsymbol{\Psi}}$ is given by

$$\hat{\boldsymbol{\Psi}} = (\boldsymbol{A'A})^{-1}\hat{\boldsymbol{S}}_{a1.b}\tilde{\boldsymbol{\beta}}_4 \tag{8.5}$$

with $\tilde{\boldsymbol{\beta}}_4 = (\tilde{\boldsymbol{v}}_{4.1}, \ldots, \tilde{\boldsymbol{v}}_{4.r})$ under the maintained hypothesis of $\mathcal{H}_1(r)$.

For testing the validity of $\mathcal{H}_4 : \boldsymbol{\alpha} = \boldsymbol{A\Psi}$ given $\mathcal{H}_1(r)$, Johansen [1991] proposed the following likelihood ratio statistic

$$-2\ln(Q; \mathcal{H}_4|\mathcal{H}_1(r)) = T\sum_{i=1}^{r}\ln\left\{\frac{(1 - \tilde{\lambda}_{4.i})}{(1 - \hat{\lambda}_i)}\right\}, \tag{8.6}$$

which is asymptotically distributed as χ^2 with $r(K - m)$ degrees of freedom. This test statistic is implemented as function `alrtest()` in the package **urca**. Therefore, if $\mathcal{H}_4$ cannot be rejected, the VECM can be reduced to an m-dimensional system by conditioning to $\Delta\boldsymbol{y}_{2t}$, where $\boldsymbol{y}_{2t}$ contains the exogenous variables.

The authors applied this test to the trade-weighted foreign wholesale price index and the three-month Eurodollar interest rate; *i.e.*, p2 and i2, respectively. The restriction matrices A_1 and A_2 are then given by

$$A_1 = \begin{bmatrix} 1 & 0 & 0 & 0 \\ 0 & 0 & 0 & 0 \\ 0 & 1 & 0 & 0 \\ 0 & 0 & 1 & 0 \\ 0 & 0 & 0 & 1 \end{bmatrix}, \quad A_2 = \begin{bmatrix} 1 & 0 & 0 & 0 \\ 0 & 1 & 0 & 0 \\ 0 & 0 & 1 & 0 \\ 0 & 0 & 0 & 1 \\ 0 & 0 & 0 & 0 \end{bmatrix}.$$

In R code 8.3, these two test statistics are calculated. The matrices A_1 and A_2 are easily set up with the matrix() function. The ca.jo object corresponding to the model $\mathcal{H}_1(r = 2)$ has been created in R code 8.1 as H1.

R Code 8.3 $\mathcal{H}_4$ model: Testing for weak exogenity

```
A1 <- matrix(c(1,0,0,0,0, 0,0,1,0,0,                            1
               0,0,0,1,0, 0,0,0,0,1),                           2
             nrow=5, ncol=4)                                    3
A2 <- matrix(c(1,0,0,0,0, 0,1,0,0,0,                            4
               0,0,1,0,0, 0,0,0,1,0),                           5
             nrow=5, ncol=4)                                    6
H41 <- summary(alrtest(z = H1, A = A1, r = 2))                  7
H42 <- summary(alrtest(z = H1, A = A2, r = 2))                  8
```

The value of the test statistic is stored in the slot teststat and its marginal level of significance, the *p*-value, in the slot pval. The restricted eigenvalues can be retrieved by object@lambda and the associated eigenvectors (*i.e.*, the cointegration relations) by object@V. The new loadings calculated as in Equation (8.5) are contained in the slot W. The results of the two tests are reported in Table 8.6. For the variable p2, the null hypothesis cannot be rejected, whereas for the interest rate variable, the result is borderline given a significance level of 10%.[4]

8.1.4 Testing Restrictions on β

In this section, three statistical tests for validating different forms of restrictions on the β matrix are discussed. The hypotheses formulated about this

---

[4] Please note that the results differ from the ones reported in Johansen and Juselius [1992]. The authors report slightly smaller values for the second eigenvalues for each test statistic. Qualitatively, the test conclusion is thereby only affected for the second hypothesis, where the authors report a value of 6.34, which clearly indicates that the three-month Eurodollar interest rate cannot be considered as weakly exogenous for β.

Table 8.6. $\mathcal{H}_4$ model: Testing for weak exogenity

Variable	Test Statistic p-value	$\tilde{\lambda}_1$	$\tilde{\lambda}_2$	$\tilde{\lambda}_3$	$\tilde{\lambda}_4$
$\mathcal{H}_{4.1}\|\mathcal{H}_1(r=2)$	0.657	0.720	0.400	0.285	0.167 0.088
$\mathcal{H}_{4.2}\|\mathcal{H}_1(r=2)$	4.384	0.112	0.387	0.256	0.194 0.086

matrix are, as for the test for the $\boldsymbol{\alpha}$ matrix, linear. Furthermore, these tests do not depend on the normalization of the cointegration relations. The first test can be used to test the validity of restrictions for all cointegration relations and is termed $\mathcal{H}_3$. This test was introduced in Johansen [1988] and applied in Johansen and Juselius [1990]. A theoretical exposition can also be found in Johansen [1991] and Johansen [1995]. In the second test, it is assumed that some r_1 of the r cointegration relations are assumed to be known and the remaining r_2 cointegration relations have to be estimated. This hypothesis is termed $\mathcal{H}_5$. Finally, in the last hypothesis, $\mathcal{H}_6$, some restrictions are placed on the r_1 cointegration relations, and the remaining r_2 ones are estimated without constraints. The last two hypotheses were introduced in Johansen and Juselius [1992]. To summarize, these three hypotheses are listed below with the dimensions of the restriction matrices and the spaces of the cointegration relations to be tested:

(i) $\mathcal{H}_3 : \boldsymbol{\beta} = \boldsymbol{H}_3\boldsymbol{\varphi}$ with $\boldsymbol{H}_3(K \times s)$, $\boldsymbol{\varphi}(s \times r)$, and $r \leq s \leq K$:
    $sp(\boldsymbol{\beta}) \subset sp(\boldsymbol{H}_3)$.
(ii) $\mathcal{H}_5 : \boldsymbol{\beta} = \boldsymbol{H}_5, \boldsymbol{\Psi}$ with $\boldsymbol{H}_5(K \times r_1)$, $\boldsymbol{\Psi}(K \times r_2)$, $r = r_1 + r_2$:
     $sp(\boldsymbol{H}_5) \subset sp(\boldsymbol{\beta})$.
(iii) $\mathcal{H}_6 : \boldsymbol{\beta} = \boldsymbol{H}_6\boldsymbol{\varphi}, \boldsymbol{\Psi}$ with $\boldsymbol{H}_6(K \times s)$, $\boldsymbol{\varphi}(s \times r_1)$, $\boldsymbol{\Psi}(K \times r_2)$, $r_1 \leq s \leq K$,
      $r = r_1 + r_2$:
      $\dim(sp(\boldsymbol{\beta}) \cap sp(\boldsymbol{H}_6)) \geq r_1$.

First, the hypothesis $\mathcal{H}_3$ is presented. Johansen showed that the estimator $\hat{\boldsymbol{\varphi}}$ under this hypothesis is the eigenvector of

$$|\lambda \boldsymbol{H}_3' \hat{\boldsymbol{S}}_{11} \boldsymbol{H}_3 - \boldsymbol{H}_3' \hat{\boldsymbol{S}}_{10} \hat{\boldsymbol{S}}_{00}^{-1} \hat{\boldsymbol{S}}_{01} \boldsymbol{H}_3|. \tag{8.7}$$

The solution to this equation gives the eigenvalues $\tilde{\lambda}_1 > \ldots > \tilde{\lambda}_s > 0$. The corresponding eigenvectors are denoted as $\tilde{\boldsymbol{V}} = (\tilde{\boldsymbol{v}}_1, \ldots, \tilde{\boldsymbol{v}}_s)$. The estimate of $\hat{\boldsymbol{\varphi}}$ is then given as $(\tilde{\boldsymbol{v}}_1, \ldots, \tilde{\boldsymbol{v}}_r)$, and therefore $\hat{\boldsymbol{\beta}} = \boldsymbol{H}_3(\tilde{\boldsymbol{v}}_1, \ldots, \tilde{\boldsymbol{v}}_r)$.

The hypothesis $\mathcal{H}_3$ given $\mathcal{H}_1(r)$ can be tested with a likelihood-ratio test defined as

$$-2\ln(Q; \mathcal{H}_3|\mathcal{H}_1(r)) = T\sum_{i=1}^{r} \ln\left\{\frac{(1 - \tilde{\lambda}_{3.i})}{(1 - \hat{\lambda}_i)}\right\}, \tag{8.8}$$

which is asymptotically distributed as χ^2 with $r(K - s)$ degrees of freedom. This test is implemented as function `blrtest()` in the contributed package **urca**.

Table 8.7. $\mathcal{H}_3$ model: Restriction in all cointegration relations

Variable	Test Statistic	p-value	$\tilde{\lambda}_1$	$\tilde{\lambda}_2$	$\tilde{\lambda}_3$	$\tilde{\lambda}_4$
$\mathcal{H}_{3.1}\|\mathcal{H}_1(r=2)$	2.761	0.599	0.386	0.278	0.090	
$\mathcal{H}_{3.2}\|\mathcal{H}_1(r=2)$	13.709	0.001	0.286	0.254	0.146	0.093

Johansen and Juselius used this test statistic for validating the purchasing power parity and for testing whether the interest rate differential enters all cointegration relations. The purchasing power parity states that the variables p_1, p_2, and e_{12} enter the cointegration relations proportionally as $(1, -1, -1)$; *i.e.*, if a stationary combination of these variables exists, then it must enter the cointegration relations as $(a_i, -a_i, -a_i, *, *)$ with $i = 1, \ldots, r$. Likewise, the restriction to test whether the interest rate differential (*i.e.*, $i_1 - i_2$) enters all cointegration relations is given by the proportional cointegration vector $(1, -1)$, and hence this can be formulated as $(*, *, *, b_i, -b_i)$ for $i = 1, \ldots, r$. Therefore, the two restriction matrices $\boldsymbol{H}_{3.1}$ and $\boldsymbol{H}_{3.2}$ are written as

$$\boldsymbol{H}_{3.1} = \begin{bmatrix} 1 & 0 & 0 \\ -1 & 0 & 0 \\ -1 & 0 & 0 \\ 0 & 1 & 0 \\ 0 & 0 & 1 \end{bmatrix}, \quad \boldsymbol{H}_{3.2} = \begin{bmatrix} 1 & 0 & 0 & 0 \\ 0 & 1 & 0 & 0 \\ 0 & 0 & 1 & 0 \\ 0 & 0 & 0 & 1 \\ 0 & 0 & 0 & -1 \end{bmatrix}.$$

The R code for conducting these two tests is displayed in R code 8.4. As in R code 8.3, as unrestricted model $\mathcal{H}_1$, the object H1 from R code 8.1 has been used.

R Code 8.4 $\mathcal{H}_3$ model: Testing for restrictions in all cointegration relations

```
H.31 <- matrix(c(1,-1,-1,0,0, 0,0,0,1,0, 0,0,0,0,1),          1
              c(5,3))                                          2
H.32 <- matrix(c(1,0,0,0,0, 0,1,0,0,0, 0,0,1,0,0,              3
              0,0,0,1,-1), c(5,4))                             4
H31 <- summary(blrtest(z = H1, H = H.31, r = 2))              5
H32 <- summary(blrtest(z = H1, H = H.32, r = 2))              6
```

The results of the test are displayed in Table 8.7. In the case of the purchasing power parity, the model hypothesis $\mathcal{H}_3$ cannot be rejected. This result mirrors closely the obtained $(a_i, -a_i, -a_i, *, *)$ relations in the estimated $\hat{\boldsymbol{\Pi}}$ matrix for two cointegration relations as provided in Table 8.5.[5]

---

[5] Incidentally, the two model hypotheses $\mathcal{H}_4$ and $\mathcal{H}_3$ for a given unrestricted model $\mathcal{H}_1(r)$ can be tested jointly, as shown in Johansen and Juselius [1990], for instance.

The authors directly tested whether $(1, -1, -1, 0, 0)'\boldsymbol{y}_t$ and $(0, 0, 0, 1, -1)'\boldsymbol{y}_t$ each constitute a stationary relation. Such hypotheses can be tested with the model $\mathcal{H}_5$, in which some cointegration relations are assumed to be known.

To test the hypothesis $\mathcal{H}_5$, the partial weighting matrix corresponding to $\boldsymbol{H}_5$ is concentrated out of the likelihood function. It is achieved by regressing $\boldsymbol{R}_{0t}$ and $\boldsymbol{R}_{1t}$ on $\boldsymbol{H}_5'\boldsymbol{R}_{1t}$ and thereby obtaining the new sets of residuals

$$\boldsymbol{R}_{0.ht} = \boldsymbol{R}_{0t} - \hat{\boldsymbol{S}}_{01}\boldsymbol{H}_5(\boldsymbol{H}_5'\hat{\boldsymbol{S}}_{11}\boldsymbol{H}_5)^{-1}\boldsymbol{H}_5\boldsymbol{R}_{1t}, \tag{8.9a}$$

$$\boldsymbol{R}_{1.ht} = \boldsymbol{R}_{1t} - \hat{\boldsymbol{S}}_{11}\boldsymbol{H}_5(\boldsymbol{H}_5'\hat{\boldsymbol{S}}_{11}\boldsymbol{H}_5)^{-1}\boldsymbol{H}_5\boldsymbol{R}_{1t}. \tag{8.9b}$$

The new product moment matrices are then calculated as

$$\hat{\boldsymbol{S}}_{ij.h} = \hat{\boldsymbol{S}}_{ij} - \hat{\boldsymbol{S}}_{i1}\boldsymbol{H}_5(\boldsymbol{H}_5'\hat{\boldsymbol{S}}_{11}\boldsymbol{H})^{-1}\boldsymbol{H}'\hat{\boldsymbol{S}}_{1j} \text{ for } i, j = 0, 1. \tag{8.10}$$

An estimate of the partially unknown cointegration relations $\hat{\boldsymbol{\Psi}}$ is obtained by solving two eigenvalue problems. First, the $K - r_1$ eigenvalues of

$$|\tau\boldsymbol{I} - \hat{\boldsymbol{S}}_{11.h}| = 0 \tag{8.11}$$

are retrieved and the auxiliary matrix $\boldsymbol{C}$ is calculated as

$$\boldsymbol{C} = (\boldsymbol{e}_1, \boldsymbol{e}_2, \ldots, \boldsymbol{e}_{K-r_1}) \begin{pmatrix} \tau_1^{-1/2} & 0 & \cdots\cdots & 0 \\ 0 & \tau_2^{-1/2} & 0 & \cdots & 0 \\ \vdots & 0 & \ddots & & \vdots \\ \vdots & \vdots & & \ddots & 0 \\ 0 & 0 & \cdots & 0 & \tau_{K-r_1}^{-1/2} \end{pmatrix}, \tag{8.12}$$

where $(\boldsymbol{e}_1, \boldsymbol{e}_2, \ldots, \boldsymbol{e}_{K-r_1})$ are the eigenvectors belonging to $\boldsymbol{\tau}$. The matrix $\boldsymbol{C}$ then enters the second eigenvalue problem,

$$|\boldsymbol{\lambda}\boldsymbol{I} - \boldsymbol{C}'\hat{\boldsymbol{S}}_{10.h}\hat{\boldsymbol{S}}_{00.h}^{-1}\hat{\boldsymbol{S}}_{01.h}\boldsymbol{C}| = 0, \tag{8.13}$$

which yields the eigenvalues $\tilde{\lambda}_1 > \ldots > \tilde{\lambda}_{K-r_1} > 0$ and eigenvectors $\tilde{\boldsymbol{V}} = (\tilde{\boldsymbol{v}}_1, \ldots, \tilde{\boldsymbol{v}}_{K-r_1})$. The partial cointegration relations are then estimated as $\hat{\boldsymbol{\Psi}} = \boldsymbol{C}(\tilde{\boldsymbol{v}}_1, \ldots, \tilde{\boldsymbol{v}}_{r_2})$, and therefore the cointegration relations are given as $\hat{\boldsymbol{\beta}} = (\boldsymbol{H}_5, \hat{\boldsymbol{\Psi}})$.

Finally, for calculating the likelihood-ratio test statistic, the eigenvalues $\boldsymbol{\rho}$ have to be extracted from

$$|\rho\boldsymbol{H}_5'\hat{\boldsymbol{S}}_{11}\boldsymbol{H}_5 - \boldsymbol{H}_5'\hat{\boldsymbol{S}}_{10}\hat{\boldsymbol{S}}_{00}^{-1}\hat{\boldsymbol{S}}_{01}\boldsymbol{H}_5| = 0 \tag{8.14}$$

with $\hat{\rho} = \hat{\rho}_1, \ldots, \hat{\rho}_{r_1}$. The test statistic is defined as

---

This combined test is implemented as the function `ablrtest()` in the contributed package **urca**. An example of its application is provided in the citation above, which is mirrored in `example(ablrtest)`.

Table 8.8. $\mathcal{H}_5$ model: Partly known cointegration relations

Variable	Test Statistic p-value		$\tilde{\lambda}_1$	$\tilde{\lambda}_2$	$\tilde{\lambda}_3$	$\tilde{\lambda}_4$
$\mathcal{H}_{5.1}\vert\mathcal{H}_1(r=2)$	14.521	0.002	0.396	0.281	0.254	0.101
$\mathcal{H}_{5.2}\vert\mathcal{H}_1(r=2)$	1.895	0.595	0.406	0.261	0.105	0.101

$$-2\ln Q(\mathcal{H}_5\vert\mathcal{H}_1(r)) = T\left\{\sum_{i=1}^{r_1}\ln(1-\hat{\rho}_i) + \sum_{i=1}^{r_2}\ln(1-\tilde{\lambda}_i) - \sum_{i=1}^{r}\ln(1-\hat{\lambda}_i)\right\},$$

(8.15)

which is asymptotically distributed as χ^2 with $(K-r)r_1$ degrees of freedom.

Johansen and Juselius applied this test statistic to validate whether the purchasing power parity or the interest rate differential form a stationary process by themselves. This test is implemented in the function bh5lrtest() contained in the contributed package **urca**. In R code 8.5, the results are replicated. The assumed-to-be-known partial cointegration matrices are set up as matrix objects H.51 and H.52, respectively. In the following lines, the test is applied to both of them.

R **Code 8.5** $\mathcal{H}_3$ model: Testing for partly known cointegration relations

```
H.51 <- c(1, -1, -1, 0, 0)                                          1
H.52 <- c(0, 0, 0, 1, -1)                                           2
H51 <- summary(bh5lrtest(z = H1, H = H.51, r = 2))                  3
H52 <- summary(bh5lrtest(z = H1, H = H.52, r = 2))                  4
```

The results are exhibited in Table 8.8. The hypothesis that the PPP relation is stationary is rejected, whereas the hypothesis that the interest differential forms a stationary process cannot be rejected.

Finally, the model hypothesis $\mathcal{H}_6 : \boldsymbol{\beta} = (\boldsymbol{H}_6\boldsymbol{\varphi}, \boldsymbol{\Psi})$ has to be discussed. Recall that this hypothesis is used for testing some restrictions placed on the first r_1 cointegration relations, and the remaining ones contained in $\boldsymbol{\Psi}$ are estimated freely. In contrast to the previous two model hypotheses, one cannot reduce this one to a simple eigenvalue problem. Johansen and Juselius [1992] proposed a simple switching algorithm instead. The algorithm is initialized by setting $\boldsymbol{\Psi} = 0$, and the eigenvalue problem

$$\vert\lambda\boldsymbol{H}_6'\hat{\boldsymbol{S}}_{11}\boldsymbol{H}_6 - \boldsymbol{H}_6'\hat{\boldsymbol{S}}_{10}\hat{\boldsymbol{S}}_{00}^{-1}\hat{\boldsymbol{S}}_{01}\boldsymbol{H}_6\vert = 0$$

(8.16)

is solved for $\boldsymbol{\varphi}$, which results in the eigenvalues $\hat{\lambda}_1 > \ldots > \hat{\lambda}_s > 0$ and the corresponding eigenvectors $(\hat{\boldsymbol{v}}_1, \ldots, \hat{\boldsymbol{v}}_s)$. The first partition of the cointegration relations (*i.e.*, the restricted ones) is therefore given by $\hat{\boldsymbol{\beta}}_1 = \boldsymbol{H}_6(\hat{\boldsymbol{v}}_1, \ldots, \hat{\boldsymbol{v}}_{r_1})$, although it is preliminary. The algorithm starts by fixing these values $\hat{\beta}_1$ and by conditioning on $\hat{\beta}_1 R_{1t}$. It leads to the eigenvalue problem

$$\frac{|\Psi'(\hat{S}_{11.\hat{\beta}_1} - \hat{S}_{10.\hat{\beta}_1}\hat{S}_{00.\hat{\beta}_1}^{-1}\hat{S}_{01.\hat{\beta}_1})\Psi|}{|\Psi'\hat{S}_{11.\hat{\beta}_1}\Psi|} \tag{8.17}$$

for Ψ, where the product moment matrices $\hat{S}_{ij.b}$ are given by

$$\hat{S}_{ij.b} = \hat{S}_{ij} - \hat{S}_{i1}\hat{\beta}_b(\hat{\beta}_b'\hat{S}_{11}\hat{\beta}_b)^{-1}\hat{\beta}_b'\hat{S}_{1j} \text{ for } i,j = 0,1 \text{ and } b = 1,2. \tag{8.18}$$

The solution to the eigenvalue problem in Equation (8.17) is given as Lemma 1 in Johansen and Juselius [1992], and an extended exposition of eigenvalues and eigenvectors is given in Appendix A.1 in Johansen [1995]. Equation (8.18) yields eigenvalues $\tilde{\lambda}_1, \ldots, \tilde{\lambda}_{K-r_1}$ and eigenvectors $(\hat{u}_1, \ldots, \hat{u}_{K-r_1})$. Hence, the second partition of cointegration relations is given as $\hat{\beta}_2 = (\hat{u}_1, \ldots, \hat{u}_{r_2})$, although it is preliminary. The second step of the algorithm consists of holding these cointegration relations fixed and conditioning on $\hat{\beta}_2 R_{1t}$. Hereby, a new estimate of $\hat{\beta}_1$ is obtained by solving

$$\frac{|\varphi' H_6'(\hat{S}_{11.\hat{\beta}_2} - \hat{S}_{10.\hat{\beta}_2}\hat{S}_{00.\hat{\beta}_2}^{-1}\hat{S}_{01.\hat{\beta}_2})H_6\varphi|}{|\varphi' H_6'\hat{S}_{11.\hat{\beta}_2}H_6\varphi|}, \tag{8.19}$$

which results in eigenvalues $\hat{\omega}_1, \ldots, \hat{\omega}_s$ and eigenvectors $(\hat{v}_1, \ldots, \hat{v}_s)$. The new estimate for β_1 is then given by $\hat{\beta}_1 = H_6(\hat{v}_1, \ldots, \hat{v}_{r_1})$. Equations (8.17) and (8.18) form the switching algorithm by consecutively calculating new sets of eigenvalues and corresponding eigenvectors until convergence is achieved; *i.e.*, the change in values from one iteration to the next is smaller than an *a priori* given convergence criterion. Alternatively, one could iterate as long as the likelihood function

$$L_{\max}^{-2/T} = \left|\hat{S}_{00.\hat{\beta}_1}\right|\prod_{i=1}^{r_2}(1-\tilde{\lambda}_i) = \left|\hat{S}_{00.\hat{\beta}_2}\right|\prod_{i=1}^{r_1}(1-\hat{\omega}_i) \tag{8.20}$$

has not achieved its maximum. Unfortunately, this algorithm does not necessarily converge to a global maximum but to a local one instead.

Finally, to calculate the likelihood-ratio test statistic, the eigenvalue problem

$$|\rho\hat{\beta}_1\hat{S}_{11}\hat{\beta}_1 - \hat{\beta}_1\hat{S}_{10}\hat{S}_{00}^{-1}\hat{S}_{01}\hat{\beta}_1| = 0 \tag{8.21}$$

has to be solved for the eigenvalues $\hat{\rho}_1, \ldots, \hat{\rho}_{r_1}$. The test statistic is then given as

$$-2\ln Q(\mathcal{H}_6|\mathcal{H}_1(r)) = T\left\{\sum_{i=1}^{r_1}\ln(1-\hat{\rho}_i) + \sum_{i=1}^{r_2}\ln(1-\tilde{\lambda}_i) - \sum_{i=1}^{r}\ln(1-\hat{\lambda}_i)\right\}, \tag{8.22}$$

Table 8.9. $\mathcal{H}_6$ model: Restrictions on r_1 cointegration relations

Variable	Test Statistic p-value	$\tilde{\lambda}_1$	$\tilde{\lambda}_2$	$\tilde{\lambda}_3$	$\tilde{\lambda}_4$
$\mathcal{H}_6\|\mathcal{H}_1(r=2)$	4.931 0.026	0.407	0.281	0.149	0.091

which is asymptotically distributed as χ^2 with $(K - s - r_2)r_1$ degrees of freedom.

This test statistic is implemented as function `bh6lrtest()` in the contributed package **urca**. Besides the $\mathcal{H}_1$ object and the restriction matrix, the total number of cointegration relations, the number of restricted relationships, the convergence value, and the maximum number of iterations enter as functional arguments. The convergence criterion is defined as the vector norm of $\tilde{\lambda}$.

Because the test result of the model hypothesis $\mathcal{H}_5$ indicated that the purchasing power parity does not hold in the strict sense, the authors applied this test to see whether a more general linear but still stationary combination of p_1, p_2, and e_{12} exists. That is, the question now is whether a more general cointegration vector of the form $(a, b, c, 0, 0)$ yields a stationary process. This restriction can be cast into the following matrix $\boldsymbol{H}_6$:

$$\boldsymbol{H}_6 = \begin{bmatrix} 1 & 0 & 0 \\ 0 & 1 & 0 \\ 0 & 0 & 1 \\ 0 & 0 & 0 \\ 0 & 0 & 0 \end{bmatrix}.$$

The application of this test is provided in R code 8.6, and its results are depicted in Table 8.9. The test statistic is not significant at the 1% level. Please note that compared with the results in Johansen and Juselius [1992], the algorithm converged to slightly different values for the second, third, and fourth eigenvalues.

R Code 8.6 $\mathcal{H}_6$ model: Testing of restrictions on r_1 cointegration relations

```
H.6 <- matrix(rbind(diag(3), c(0, 0, 0), c(0, 0, 0)),      1
              nrow=5, ncol=3)                               2
H6 <- summary(bh6lrtest(z = H1, H = H.6,                   3
                  r = 2, r1 = 1))                           4
```

8.2 VECM and Structural Shift

In Section 6.1, the implications for the statistical inference of unit root tests in light of structural breaks have been discussed. The pitfalls of falsely concluding non-stationarity in the data can also be encountered in the case of VECM. The flip side would be a wrongly accepted cointegration relation, where some or all underlying series behave like an AR(1)-process with a structural break. Lütkepohl, Saikkonen and Trenkler [2004] proposed a procedure for estimating a VECM in which the structural shift is a simple shift in the level of the process and the break date is estimated first. Next, the deterministic part, including the size of the shift, is estimated, and the data are adjusted accordingly. Finally, a Johansen-type test for determining the cointegration rank can be applied to these adjusted series.

Lütkepohl et al. assume that the $(K \times 1)$ vector process $\{y_t\}$ is generated by a constant, a linear trend, and level shift terms

$$y_t = \mu_0 + \mu_1 t + \delta d_{t\tau} + x_t, \tag{8.23}$$

where $d_{t\tau}$ is a dummy variable defined by $d_{t\tau} = 0$ for $t < \tau$ and $d_{t\tau} = 1$ for $t \geq \tau$. The shift assumes that the shift point τ is unknown and is expressed as a fixed fraction of the sample size,

$$\tau = [T\lambda] \text{ with } 0 < \underline{\lambda} \leq \lambda \leq \overline{\lambda} < 1, \tag{8.24}$$

where $\underline{\lambda}$ and $\overline{\lambda}$ define real numbers and $[\cdot]$ defines the integer part. The meaning of Equation (8.24) is that the shift might occur neither at the very beginning nor at the very end of the sample. Furthermore, it is assumed that the process $\{x_t\}$ can be represented as a VAR(p) and that the components are at most $I(1)$ and cointegrated with rank r.

The estimation of the break point is based on the regressions

$$y_t = \nu_0 + \nu_1 t + \delta d_{t\tau} + A_1 y_{t-1} + \ldots + A_p y_{t-p} + \varepsilon_{t\tau} \text{ for } t = p+1, \ldots, T, \tag{8.25}$$

where A_i with $i = 1, \ldots, p$ assign the $(K \times K)$ coefficient matrices and ε_t is the spherical K-dimensional error process. It should be noted that other exogenous regressors, like seasonal dummy variables, can also be included in Equation (8.25).

The estimator for the break point $\hat{\tau}$ is then defined as

$$\hat{\tau} = \underset{\tau \in \mathfrak{I}}{\arg\min} \det \left(\sum_{t=p+1}^{T} \hat{\varepsilon}_{t\tau} \hat{\varepsilon}'_{t\tau} \right), \tag{8.26}$$

where $\mathfrak{I} = [T\underline{\lambda}, T\overline{\lambda}]$ and $\hat{\varepsilon}_{t\tau}$ are the least-squares residuals of Equation (8.25). The integer count of the interval $\mathfrak{I} = [T\underline{\lambda}, T\overline{\lambda}]$ determines how many regressions have to be run with the corresponding step dummy variables $d_{t\tau}$ and how many times the determinant of the product moment matrices of $\hat{\varepsilon}_{t\tau}$ have to be calculated. The minimal one is the one that selects the most likely break point.

Once the break point $\hat{\tau}$ is estimated, the data are adjusted according to

$$\hat{x}_t = y_t - \hat{\mu}_0 - \hat{\mu}_1 t - \hat{\delta} d_{t\hat{\tau}}. \tag{8.27}$$

This method is included as function `cajolst()` in the contributed package **urca**. By applying this function, an object of class `ca.jo` is generated. The adjusted series are in the slot x, and the estimate of the break point is stored in the slot bp. Instead of using the test statistic as proposed in Lütkepohl et al. [2004] with critical values provided in Lütkepohl and Saikkonen [2000], the test statistic

$$LR(r) = T \sum_{j=r+1}^{N} \ln(1 + \hat{\lambda}_j) \tag{8.28}$$

has been implemented with critical values from Trenkler [2003] in the function `cajolst()`. The advantage is that, in the latter source, the critical values are provided more extensively and precisely.

In R code 8.7, this method has been applied to estimate a money demand function for Denmark as in Johansen and Juselius [1990]. For a better comparison, the results for the non-adjusted data are also given in Table 8.10.

R Code 8.7 $\mathcal{H}_1$ model: Inference on cointegration rank for Danish money demand function allowing for structural shift

```
data(denmark)                                                    1
sjd <- denmark[, c("LRM", "LRY", "IBO", "IDE")]                  2
sjd.vecm <- summary(ca.jo(sjd, ecdet = "const",                  3
                    type = "eigen",                              4
                    K = 2,                                       5
                    spec = "longrun",                            6
                    season = 4))                                 7
lue.vecm <- summary(cajolst(sjd, season=4))                      8
```

For the non-adjusted data, the hypothesis of one cointegration relation cannot be rejected for a significance level of 5%. If one allows for a structural shift in the data, however, one cannot reject the hypothesis of no cointegration as indicated by the results in Table 8.11. The shift occurred most likely in 1975:Q4. Therefore, a VAR in differences with an intervention dummy for that period might be a more suitable model to describe the data-generating process.

Table 8.10. Money demand function for Denmark: Maximal eigenvalue statistic, non-adjusted data

Rank	Test Statistic	10%	5%	1%
$r <= 3$	2.35	7.52	9.24	12.97
$r <= 2$	6.34	13.75	15.67	20.20
$r <= 1$	10.36	19.77	22.00	26.81
$r = 0$	30.09	25.56	28.14	33.24

Table 8.11. Money demand function for Denmark: Trace statistic, allowing for structural shift

Rank	Test Statistic	10%	5%	1%
$r <= 3$	3.15	5.42	6.79	10.04
$r <= 2$	11.62	13.78	15.83	19.85
$r <= 1$	24.33	25.93	28.45	33.76
$r = 0$	42.95	42.08	45.20	51.60

8.3 The Structural Vector Error-Correction Model

Reconsider the VECM from Equation (4.9) on page 80. It is possible to apply the same reasoning for SVAR-models as outlined in Section 2.3 to VECMs, in particular when the equivalent level-VAR representation of the VECM is used. However, the information contained in the cointegration properties of the variables is therefore not used for identifying restrictions on the structural shocks. Hence, typically a B-type model is assumed, whence an SVEC-model is specified and estimated.

$$\Delta y_t = \alpha \beta' y_{t-1} + \Gamma_1 \Delta y_{t-1} + \ldots + \Gamma_{p-1} y_{t-p+1} + B\varepsilon_t, \tag{8.29}$$

where $u_t = B\varepsilon_t$ and $\varepsilon_t \sim N(0, I_K)$. In order to exploit this information, one considers the Beveridge-Nelson moving average representation of the variables y_t if they adhere to the VECM process as in Equation (3.4):

$$y_t = \Xi \sum_{i=1}^{t} u_i + \sum_{j=0}^{\infty} \Xi_j^* u_{t-j} + y_0^*. \tag{8.30}$$

The variables contained in y_t can be decomposed into a part that is integrated of order one and a part that is integrated of order zero. The first term on the right-hand side of Equation (8.30) is referred to as the "common trend" of the system, and this term drives the system y_t. The middle term is integrated of order zero, and it is assumed that the infinite sum is bounded; *i.e.*, Ξ_j^* converge to zero as $j \to \infty$. The initial values are captured by y_0^*. For the modeling of SVEC, interest centers on the common trends, in which the long-run effects of shocks are captured. The matrix $\varXi$ is of reduced rank $K - r$, where r is the count of stationary cointegration relationships. The matrix is defined as

$$\varXi = \beta_\perp \left[\alpha'_\perp \left(I_K - \sum_{i=1}^{p-1} \varGamma_i \right) \beta_\perp \right]^{-1} \alpha'_\perp. \tag{8.31}$$

Because of its reduced rank, only $K - r$ common trends drive the system. Therefore, by knowing the rank of $\varPi$, one can then conclude that at most r of the structural errors can have a transitory effect. This implies that at most r columns of $\varXi$ can be set to zero. One can combine the Beveridge-Nelson decomposition with the relationship between the VECM error terms and the structural innovations. The common trends term is then $\varXi B \sum_{t=1}^{\infty} \varepsilon_t$, and the long-run effects of the structural innovations are captured by the matrix $\varXi B$. The contemporaneous effects of the structural errors are contained in the matrix B. As in the case of SVAR-models of type B, one needs for local, just-identified SVEC-models $\frac{1}{2} K(K-1)$ restrictions. The cointegration structure of the model provides $r(K-r)$ restrictions on the long-run matrix. The remaining restrictions can be placed on either matrix, where at least $r(r - 1)/2$ of them must be imposed directly on the contemporaneous matrix B.

We now specify and estimate an SVEC-model and thereby replicate the results in Breitung, Brüggemann and Lütkepohl [2004]. The authors investigated the Canadian labor market by using a model that was proposed by Jaconson, Vredin and Warne [1997]. The model consists of four equations: a production function, a labor demand function, a labor supply function, and a wage-setting relation. The output equation is specified as

$$gdp_t = \rho e_t + \theta_{1,t}, \tag{8.32}$$

the output is dependent on employment, and the coefficient ρ measures returns to scale. It is further assumed that advances in technology can be represented by the quantity $\theta_{1,t}$, which follows a random walk; i.e., $\theta_{1,t} = \theta_{1,t-1} + \varepsilon_1^{gdp}$ and ε_1^{gdp} is white noise. The labor demand is a function of output and real wages,

$$e_t = \lambda gdp_t - \nu(w - p)_t + \theta_{2,t}, \tag{8.33}$$

where it is assumed that $\theta_{2,t} = \phi_D \theta_{2,t-1} + \varepsilon_1^e$ is an AR(1)-process. The labor demand is therefore stationary if the absolute value of the autoregressive coefficient is less than 1. The labor supply is dependent on the real wage and a trend. For the latter, it is assumed that this trend follows a random walk

$$l_t = \pi(w - p)_t + \theta_{3,t}, \tag{8.34}$$

with $\theta_{3,t} = \theta_{3,t-1} + \varepsilon_t^s$. Finally, real wages are a function of labor productivity and unemployment, and its trend component is modeled, like the labor demand equation, as an AR(1)-process that can be non-stationary,

$$(w - p)_t = \delta(gdp - e)_t - \gamma(l - e)_t + \theta_{4,t}, \tag{8.35}$$

with $\theta_{4,t} = \phi_u \theta_{4,t-1} + \varepsilon_t^w$. The derivation of the model's solution in terms of the trend variables $\theta_{i,t}$ for $i = 1, \ldots, 4$ as right-hand-side arguments is left to the reader (see Exercise 4).

Breitung et al. [2004] utilized the following series: labor productivity defined as the log difference between GDP and employment, the log of employment, the unemployment rate, and real wages defined as the log of the real wage index. These series are signified by "prod," "e," "U," and "rw," respectively. The data are taken from the OECD database and span from the first quarter 1980 until the fourth quarter 2004.

R Code 8.8 Canadian data set: Preliminary analysis

```
library ( vars )                                                1
data ( Canada )                                                 2
summary ( Canada )                                              3
plot ( Canada ,   nc  =  2)                                     4
```

A preliminary data analysis is conducted by displaying the summary statistics of the series involved as well as the corresponding time series plots (see Figure 8.3). In the next step, Breitung et al. conducted unit root tests by applying the ADF test regressions to the series. The R code for the tests conducted is given in R code 8.9, and the results are reported in Table 8.12.

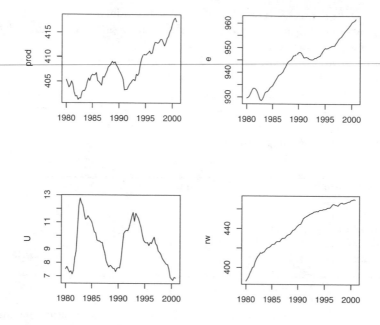

Fig. 8.3. Canadian data set: Time series plots

R **Code 8.9** Canadian data set: ADF-test regressions

```
summary(ur.df(Canada[, "prod"],                           1
          type = "trend", lags = 2))                      2
summary(ur.df(diff(Canada[, "prod"]),                     3
          type = "drift", lags = 1))                      4
summary(ur.df(Canada[, "e"],                              5
          type = "trend", lags = 2))                      6
summary(ur.df(diff(Canada[, "e"]),                        7
          type = "drift", lags = 1))                      8
summary(ur.df(Canada[, "U"],                              9
          type = "drift", lags = 1))                      10
summary(ur.df(diff(Canada[, "U"]),                        11
          type = "none", lags = 0))                       12
summary(ur.df(Canada[, "rw"],                             13
          type = "trend", lags = 4))                      14
summary(ur.df(diff(Canada[, "rw"]),                       15
          type = "drift", lags = 3))                       16
summary(ur.df(diff(Canada[, "rw"]),                       17
          type = "drift", lags = 0))                       18
```

Table 8.12. ADF tests for Canadian data

Variable	Deterministic Terms	Lags	Test Value	Critical Values		
				1%	5%	10%
$prod$	constant, trend	2	−1.99	−4.04	−3.45	−3.15
$\Delta prod$	constant	1	−5.16	−3.51	−2.89	−2.58
e	constant, trend	2	−1.91	−4.04	−3.45	−3.15
Δe	constant	1	−4.51	−3.51	−2.89	−2.58
U	constant	1	−2.22	−3.51	−2.89	−2.58
ΔU		0	−4.75	−2.6	−1.95	−1.61
rw	constant, trend	4	−2.06	−4.04	−3.45	−3.15
Δrw	constant	3	−2.62	−3.51	−2.89	−2.58
Δrw	constant	0	−5.6	−3.51	−2.89	−2.58

Table 8.13. Canada VAR: Lag-order selection

Lag Order	AIC(n)	HQ(n)	SC(n)	FPE(n)
$p = 1$	−6.2726	−5.9784	−5.5366	0.0019
$p = 2$	−6.6367	−6.1464	−5.4100	0.0013
$p = 3$	−6.7712	−6.0848	−5.0538	0.0012
$p = 4$	−6.6346	−5.7522	−4.4265	0.0014
$p = 5$	−6.3981	−5.3196	−3.6994	0.0018
$p = 6$	−6.3077	−5.0331	−3.1183	0.0020
$p = 7$	−6.0707	−4.6000	−2.3906	0.0028
$p = 8$	−6.0616	−4.3947	−1.8908	0.0031

It can be concluded that all time series are integrated of order one. Please note that the critical values reported differ slightly from the ones that are reported in Breitung et al. [2004]. The authors utilized the software JMULTI, in which the critical values of Davidson and MacKinnon [1993] are used, whereas in the function ur.df() the critical values are taken from Dickey and Fuller [1981] and Hamilton [1994].

In an ensuing step, the authors determined an optimal lag length for an unrestricted VAR for a maximal lag length of 8. This can be accomplished swiftly with the function VARselect(), as evidenced in R code 8.10. The results are reported in Table 8.13.

R Code 8.10 Canada VAR: Lag-order selection

```
VARselect(Canada, lag.max = 8, type = "both")                          1
```

According to the AIC and FPE, the optimal lag number is $p = 3$, whereas the HQ criterion indicates $p = 2$ and the SC criterion indicates an optimal lag length of $p = 1$. Breitung et al. estimated for all three lag orders a VAR including a constant and a trend as deterministic regressors and conducted diagnostic tests with respect to the residuals. In R code 8.11, the relevant commands are exhibited. First, the variables have to be reordered in the same sequence as in Breitung et al. [2004]. This step is necessary because otherwise the results of the multivariate Jarque-Bera test, in which a Choleski decomposition is employed, would differ slightly from those reported in Breitung et al. [2004].

R Code 8.11 Diagnostic tests for VAR(p) specifications for Canadian data

```
Canada <- Canada[, c("prod", "e", "U", "rw")]                1
p1ct <- VAR(Canada, p = 1, type = "both")                    2
p2ct <- VAR(Canada, p = 2, type = "both")                    3
p3ct <- VAR(Canada, p = 3, type = "both")                    4
## Serial                                                    5
serial.test(p3ct, lags.pt = 16,                              6
            type = "PT.asymptotic")                          7
serial.test(p2ct, lags.pt = 16,                              8
            type = "PT.asymptotic")                          9
serial.test(p1ct, lags.pt = 16,                             10
            type = "PT.asymptotic")                         11
serial.test(p3ct, lags.pt = 16,                             12
            type = "PT.adjusted")                           13
serial.test(p2ct, lags.pt = 16,                             14
            type = "PT.adjusted")                           15
serial.test(p1ct, lags.pt = 16,                             16
            type = "PT.adjusted")                           17
## JB                                                       18
normality.test(p3ct)                                        19
normality.test(p2ct)                                        20
normality.test(p1ct)                                        21
## ARCH                                                     22
arch.test(p3ct, lags.multi = 5)                             23
arch.test(p2ct, lags.multi = 5)                             24
arch.test(p1ct, lags.multi = 5)                             25
## Stability (Recursive CUSUM)                              26
plot(stability(p3ct), nc = 2)                               27
plot(stability(p2ct), nc = 2)                               28
plot(stability(p1ct), nc = 2)                               29
```

Given the diagnostic test results, Breitung et al. concluded that a VAR(1)-specification might be too restrictive. The graphical results of the OLS-

Table 8.14. Diagnostic tests for VAR(p) specifications for Canadian data

Lag order	Q_{16}	p-value	Q_{16}^*	p-value	JB_4	p-value	$MARCH_5$	p-value
$p = 3$	173.97	0.96	198.04	0.68	9.66	0.29	512.04	0.35
$p = 2$	209.74	0.74	236.08	0.28	2.29	0.97	528.14	0.19
$p = 1$	233.50	0.61	256.88	0.22	9.92	0.27	570.14	0.02

CUSUM test are exhibited in Figures 8.4–8.6. They argued further that although some of the stability tests do indicate deviations from parameter constancy, the time-invariant specifications of the VAR(2) and VAR(3) models will be maintained as tentative candidates for the following cointegration analysis.

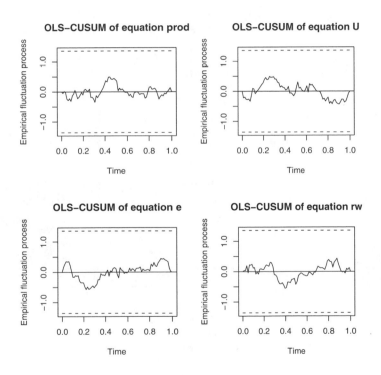

Fig. 8.4. OLS-CUSUM test of VAR(3)

The authors estimated a VECM where a deterministic trend has been included in the cointegration relation. The estimation of these models as well as the statistical inference with respect to the cointegration rank can be swiftly accomplished with the function `ca.jo()`. Although the following R code exam-

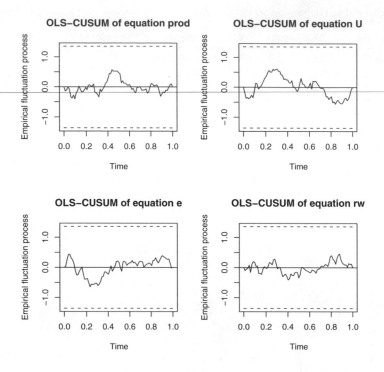

Fig. 8.5. OLS-CUSUM test of VAR(2)

ples use functions contained in the package **urca**, it is beneficial to reproduce these results for two reasons: the interplay between the functions contained in the packages **urca** and **vars** is exhibited and provides an understanding of the SVEC specification that follows.

R Code 8.12 Johansen cointegration tests for Canadian system

```
summary(ca.jo(Canada, type = "trace",           1
              ecdet = "trend", K = 3,            2
              spec = "transitory"))              3
summary(ca.jo(Canada, type = "trace",           4
              ecdet = "trend", K = 2,            5
              spec = "transitory"))              6
```

These results indicate one cointegration relationship. The reported critical values differ slightly from the ones that are reported in Table 4.3 of Breitung et al. [2004]. These authors used the values that are contained in Johansen [1995], whereas the values from Osterwald-Lenum [1992] are used in the func-

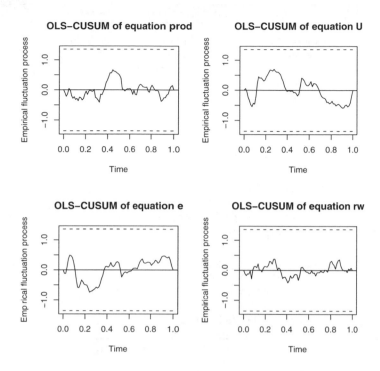

Fig. 8.6. OLS-CUSUM test of VAR(1)

Table 8.15. Johansen cointegration tests for Canadian system

$\mathcal{H}_0$	Test Statistics		Critical Values		
	$p = 3$	$p = 2$	90%	95%	99%
$r = 0$	84.92	86.12	59.14	62.99	70.05
$r = 1$	36.42	37.33	39.06	42.44	48.45
$r = 2$	18.72	15.65	22.76	25.32	30.45
$r = 3$	3.85	4.10	10.49	12.25	16.26

tion `ca.jo()`. In R code 8.13, the VECM is reestimated with this restriction and a normalization of the long-run relationship with respect to real wages. The results are shown in Table 8.16.

R Code 8.13 VECM with $r = 1$ and normalization with respect to real wages

```
vecm <- ca.jo(Canada[, c("rw", "prod", "e", "U")],      1
              type = "trace", ecdet = "trend",          2
              K = 3, spec = "transitory")               3
vecm.r1 <- cajorls(vecm, r = 1)                         4
alpha <- coef(vecm.r1$rlm)[1, ]                         5
beta <- vecm.r1$beta                                    6
resids <- resid(vecm.r1$rlm)                            7
N <- nrow(resids)                                       8
sigma <- crossprod(resids) / N                          9
## t-stats for alpha                                    10
alpha.se <- sqrt(solve(crossprod(                       11
              cbind(vecm@ZK %*% beta, vecm@Z1)))        12
              [1, 1]* diag(sigma))                      13
alpha.t <- alpha / alpha.se                             14
## t-stats for beta                                     15
beta.se <- sqrt(diag(kronecker(solve(                   16
              crossprod(vecm@RK[, -1])),                17
              solve(t(alpha) %*% solve(sigma)           18
              %*% alpha))))                             19
beta.t <- c(NA, beta[-1] / beta.se)                     20
```

Table 8.16. Cointegration vector and loading parameters

Vector	prod	e	U	rw	trend
$\hat{\beta}'$	0.545	−0.013	1.727	1.000	−0.709
	(0.90)	(−0.02)	(1.19)		(−2.57)
$\hat{\alpha}'$	−0.012	−0.016	−0.009	−0.085	
	(−0.92)	(−2.16)	(−1.49)	(−5.71)	

Note: t statistics in parentheses.

For a just identified SVEC-model of type B, one needs $\frac{1}{2}K(K-1) = 6$ linear independent restrictions. It is further reasoned from the Beveridge-Nelson decomposition that there are $k^* = r(K-r) = 3$ shocks with permanent effects and only one shock that exerts a temporary effect, due to $r = 1$. Because the cointegration relation is interpreted as a stationary wage-setting relation, the temporary shock is associated with the wage shock variable. Hence, the four entries in the last column of the long-run impact matrix ΞB are set to zero. Because this matrix is of reduced rank, only $k^* r = 3$ linear independent restrictions are imposed. It is therefore necessary to set $\frac{1}{2}k^*(k^*-1) = 3$ additional elements to zero. Breitung et al. assumed constant-scale returns and that therefore productivity is only driven by output shocks. This reasoning implies zero coefficients in the first row of the long-run matrix for the variables employment, unemployment, and real wages, and hence the elements $\Xi B_{1,j}$ for $j = 2, 3, 4$ are set to zero. Because $\Xi B_{1,4}$ has already been set to zero, only two additional restrictions have been added. The last restriction is imposed on the element $B_{4,2}$. The authors assumed that labor demand shocks do not exert an immediate effect on real wages.

In R code 8.14, the matrix objects LR and SR are set up accordingly and the just-identified SVEC is estimated with function SVEC(). In the call to the function SVEC(), the argument **boot = TRUE** has been employed such that bootstrapped standard errors and hence t statistics can be computed for the structural long-run and short-run coefficients.

R Code 8.14 Estimation of SVEC with bootstrapped t statistics

```
vecm <- ca.jo(Canada[, c("prod", "e", "U", "rw")],          1
              type = "trace", ecdet = "trend",              2
              K = 3, spec = "transitory")                   3
SR <- matrix(NA, nrow = 4, ncol = 4)                        4
SR[4, 2] <- 0                                               5
SR                                                          6
LR <- matrix(NA, nrow = 4, ncol = 4)                        7
LR[1, 2:4] <- 0                                             8
LR[2:4, 4] <- 0                                             9
LR                                                          10
svec <- SVEC(vecm, LR = LR, SR = SR, r = 1,                 11
             lrtest = FALSE,                                12
             boot = TRUE, runs = 100)                       13
svec                                                        14
svec$SR / svec$SRse                                         15
svec$LR / svec$LRse                                         16
```

The results are shown in Tables 8.17 and 8.18. The values of the t statistics differ slightly from those reported in Breitung et al. [2004], which can be attributed to sampling. In R code 8.14, only 100 runs have been executed, whereas in Breitung et al. [2004] 2000 repetitions were used.

Breitung et al. investigated further if labor supply shocks have no long-run impact on unemployment. This hypothesis is mirrored by setting $\Xi B_{3,3} = 0$. Because one more zero restriction has been added to the long-run impact matrix, the SVEC-model is now overidentified. The validity of this over-identification restriction can be tested with an LR test. In R code 8.15, first the additional restriction is set and then the SVEC is reestimated. The result of the LR test is contained in the returned list as named element LRover.

Table 8.17. Estimated coefficients of the contemporaneous impact matrix

Equation	ε_t^{gdp}	$\varepsilon_t^{Labor^d}$	$\varepsilon_t^{Labor^s}$	ε_t^{wage}
Output	0.58	0.07	−0.15	0.07
	(5.53)	(0.55)	(−0.64)	(1.02)
Labor demand	−0.12	0.26	−0.16	0.09
	(−1.68)	(4.31)	(−0.91)	(2.49)
Unemployment	0.03	−0.27	0.01	0.05
	(0.45)	(−5.88)	(0.09)	(1.54)
Real wages	0.11	0	0.48	0.49
	(0.73)		(0.74)	(6.11)

Note: t statistics in parentheses.

Table 8.18. Estimated coefficients of the long-run impact matrix

Equation	ε_t^{gdp}	$\varepsilon_t^{Labor^d}$	$\varepsilon_t^{Labor^s}$	ε_t^{wage}
Output	0.79	0	0	0
	(4.78)			
Labor demand	0.2	0.58	−0.49	0
	(0.88)	(2.85)	(−0.86)	
Unemployment	−0.16	−0.34	0.14	0
	(−1.49)	(−3.37)	(0.92)	
Real wages	−0.15	0.6	−0.25	0
	(−0.85)	(3.37)	(−0.92)	

Note: t statistics in parentheses.

R Code 8.15 SVEC: Overidentification test

```
LR[3, 3] <- 0                                              1
LR                                                         2
svec.oi <- SVEC(vecm, LR = LR, SR = SR, r = 1,             3
            lrtest = TRUE, boot = FALSE)                   4
svec.oi <- update(svec, LR = LR, lrtest = TRUE,            5
            boot = FALSE)                                  6
svec.oi$LRover                                             7
```

The value of the test statistic is 6.07, and the p-value of this $\chi^2(1)$-distributed variable is 0.014. Therefore, the null hypothesis that shocks to the labor supply do not exert a long-run effect on unemployment has to be rejected for a significance level of 5%.

In order to investigate the dynamic effects on unemployment, the authors applied an impulse response analysis. The impulse response analysis shows the effects of the different shocks (*i.e.*, output, labor demand, labor supply, and wage) to unemployment. In R code 8.16, the `irf` method for objects with class attribute `svecest` is employed and the argument `boot = TRUE` has been set such that confidence bands around the impulse response trajectories can be calculated. The outcome of the IRA is exhibited in Figure 8.7.

R Code 8.16 SVEC: Impulse response analysis

```
svec.irf <- irf(svec, response = "U",                    1
                n.ahead = 48, boot = TRUE)               2
svec.irf                                                 3
plot(svec.irf)                                           4
```

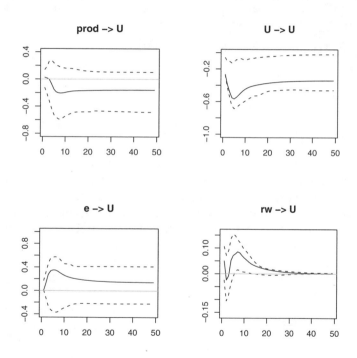

Fig. 8.7. Responses of unemployment to economic shocks with a 95% bootstrap confidence interval

In the final step, a forecast error variance decomposition is conducted with respect to unemployment. This is achieved by applying the `fevd` method to the object with class attribute `svecest`. The FEVD is swiftly computed for

Table 8.19. Forecast error variance decomposition of Canadian unemployment

Period	ε_t^{gdp}	$\varepsilon_t^{Labor^d}$	$\varepsilon_t^{Labor^s}$	ε_t^{wage}
1	0.01	0.96	0.00	0.03
4	0.01	0.78	0.21	0.01
8	0.05	0.69	0.24	0.01
12	0.08	0.68	0.23	0.01
24	0.10	0.69	0.21	0.01
48	0.12	0.70	0.18	0.00

the unemployment, as shown in R code 8.17. The authors report only the values for selected quarters. These results are displayed in Table 8.19.

R Code 8.17 Forecast error variance decomposition of Canadian unemployment

```
fevd.U <- fevd(svec, n.ahead = 48)$U                                    1
```

Summary

In this last chapter of the book, likelihood-based inference in cointegrated vector autoregressive models has been presented. It has been shown how to determine the cointegration rank and, depending on that outcome, how to specify and test the validity of restrictions placed on the cointegrating and the weighting matrices. This methodology offers the researcher a powerful tool to investigate the relationships in a system of cointegrated variables more thoroughly compared with the single-equation methods presented in Chapter 7. Furthermore, it has been shown how one can employ this methodology in light of a structural shift at an unknown point in time. The chapter concluded with an exposition of the SVEC analysis applied to a macroeconomic data set for Canada.

Exercises

1. Consider the data sets **finland** and **denmark** in the contributed package **urca**. Specify for each country a VECM that mirrors a real money demand function.

2. Reconsider the data sets **Raotbl1** and **Raotbl2** in the contributed package **urca**. Now specify a VECM for each monetary aggregate, and compare your findings with the results from Exercise 1 in Chapter 7.

3. Reconsider the data set `Raotbl6` in the contributed package **urca**. Now specify a VECM Phillips-curve model as in Mehra [1994]. Discuss your findings compared with your results from Exercise 2 in Chapter 7.

4. Solve the labor market model shown by Equations (8.32)–(8.35) for its endogenous variables $(gdp - e)_t$, e_t, $(l - e)_t$, and $(w - p)_t$ in terms of $\theta_{i,t}$ for $i = 1, \ldots, 4$.

5. Reconsider the SVEC-model for the Canadian data set. Reestimate the VECM, and conduct the analysis for a cointegration rank of $r = 2$ instead of $r = 1$.

9

Appendix

9.1 Time Series Data

There are several possibilities for dealing with time series data in R. First, the class ts in the base distribution is well suited for handling regularly spaced time series data. In R code 9.1, it is shown how to assign the range and the frequency to the data frame finland contained in the package **urca**. Objects of class ts own a time series property that can be shown by the function tsp(). The time component of an object of class ts can be retrieved with the function time(). Finally, subsetting a time series object to a narrower sample range is accomplished by using the window() function.

R Code 9.1 Time series objects of class ts

```
## time series handling in R                              1
library(urca)                                            2
data(finland)                                            3
str(finland)                                             4
## utilization of time series class 'ts'                 5
## in base package                                       6
fin.ts <- ts(finland, start=c(1958, 2),                 7
            end=c(1984, 3), frequency=4)                 8
str(fin.ts)                                              9
## time series properties of fin.ts                      10
tsp(fin.ts)                                              11
time(fin.ts)[1:10]                                       12
## Creating a subsample                                  13
finsub.ts <- window(fin.ts, start=c(1960, 2),           14
            end=c(1979, 3))                              15
tsp(finsub.ts)                                           16
```

Second, mostly encountered in financial econometric applications is the case of series irregularly spaced with respect to time. Four contributed packages exist in R that particularly address this issue: **fBasics**, **its**, **tseries**, and **zoo**. Although these packages differ in how certain functionalities and classes are defined, building unions and intersections and the merging of objects can be achieved with all of them, although the package **its** is the most mature. The functions `its()` and `timeSeries()` in the packages **its** and **fBasics** have been implemented as S4 classes, whereas the functions `irts()` and `zoo()` in the packages **tseries** and **zoo** are S3 classes for irregularly spaced observations. The advantage of **zoo** compared with the other functionalities is that time information can be of almost any class, whereas in the other implementations it needs to be of class `POSIXct`. The handling of irregular time series in the package **fBasics** resembles that for the **finmetrics** package of S-PLUS. Further details about date-time classes are provided in the RNews articles by Ripley and Hornik [2001] and Grothendieck and Petzoldt [2004].

9.2 Technicalities

This book was typeset in LaTeX. Text editor `Emacs/ESS` has been used. The indices were generated with the program `makeindex` and the bibliography with `BiBTeX`. The flow chart (see Figure 3.3) was produced with the program `flow`. The following LaTeX packages have been used: `amsmath`, `amssymb`, `array`, `bm`, `booktabs`, `float`, `graphicx`, `index`, `listings`, `multicol`, `paralist`, and `sweave`.

All R code examples have been processed as `Sweave` files. Therefore, the proper working of the R commands is guaranteed. Where possible, the results are exhibited as tables by making use of the function `latex()` contained in the contributed package **Hmisc**. The examples have been processed under R version 2.6.2 on an i486 PC with Linux as the operating system and kernel 2.6.22-14-generic. Linux is a registered trademark of Linus Torvalds (Helsinki, Finland), the original author of the Linux kernel. All contributed packages were updated before publication and are listed in Table 9.1.

9.3 CRAN Packages Used

Table 9.1. Overview of packages used

Name	Title	Version	Date
car	Companion to Applied Regression	1.2-7	2007-10-27
chron	Chronological objects which can handle dates and times	2.3-22	2008-03-04
dse1	Dynamic Systems Estimation (time series package)	2007.11-1	2007-11-01
dyn	Time Series Regression	0.2-6	2005-06-15
dynlm	Dynamic Linear Regression	0.2-0	2008-01-26
fArma	Rmetrics—ARMA Time Series Modelling	260.72	2007
fBasics	Rmetrics—Markets and Basic Statistics	260.72	2007
forecast	Forecasting functions for time series	1.11	2008-02-08
fracdiff	Fractionally differenced ARIMA aka ARFIMA(p,d,q) models	1.3-1	2006-09-07
fUnitRoots	Rmetrics—Trends and Unit Roots	260.72	2007
Hmisc	Harrell Miscellaneous	3.4-3	2007-10-31
lmtest	Testing Linear Regression Models	0.9-21	2007-07-26
mAr	Multivariate AutoRegressive analysis	1.1-1	
Rcmdr	R Commander	1.3-12	2008-01-24
strucchange	Testing, Monitoring and Dating Structural Changes	1.3-2	2007-04-13
tseries	Time series analysis and computational finance	0.10-14	2008-02-22
urca	Unit root and cointegration tests for time series data	1.1-6	2007-11-01
uroot	Unit Root Tests and Graphics for Seasonal Time Series	1.4	2005-10-10
vars	VAR Modelling	1.3-7	2008-02-12
zoo	Z's ordered observations	1.5-0	2008-03-14

10

Abbreviations, Nomenclature, and Symbols

Abbreviations:

ACF	autocorrelation function
ADF	augmented Dickey-Fuller
ADL	autoregressive distributed lag
AIC	Akaike information criteria
AR	autoregression
ARIMA	autoregressive integrated moving average
ARFIMA	autoregressive fractionally integrated moving average
ARMA	autoregressive moving average
BIC	Bayesian information criteria
CI(d, b)	cointegrated of order d, b
CRDW	cointegrating regression Durbin-Watson statistic
DF	Dickey-Fuller
DGP	data-generating process
ECM	error-correction model/mechanism
ERS	Elliott, Rothenberg, and Stock
FEVD	Forecast error variance decomposition
GNP	gross national product
HEGY	Hylleberg, Engle, Granger, and Yoo
I(d)	integrated of order d
i.d.	independently distributed
i.i.d.	independently and identically distributed
IRF	impulse response function
JB	Jarque-Bera test
LB	Ljung-Box Portmanteau test
LM	Lagrange multiplier
KPSS	Kwiatkowski, Phillips, Schmidt, and Shin

MA	moving average
NI	near integrated
OLS	ordinary least-squares
PACF	partial autocorrelation function
PP	Phillips and Perron
SC	Schwarz criteria
SI	seasonally integrated
SP	Schmidt and Phillips
SVAR	structural vector autoregressive model
SVECM	structural vector error-correction model
T	sample size or last observation in a time series
VAR	vector autoregression
var	variance
VECM	vector error-correction model
ZA	Zivot and Andrews

Nomenclature:

Bold lowercase: $\boldsymbol{y}, \boldsymbol{\alpha}$	vectors
Bold uppercase: $\boldsymbol{Y}, \boldsymbol{\Gamma}$	matrices
Greek letters: α, β, γ	population values (parameters)
Greek letters with ˆ or ˜	sample values (estimates or estimators)
Y, y	endogenous variables
X, x, Z, z	exogenous or predetermined variables
L	Lag operator, defined as $Lx_t = x_{t-1}$
Δ	first-difference operator: $\Delta x_t = x_t - x_{t-1}$

Symbols:

$\perp$	orthogonality sign
$\cap$	intersection
$\in$	set membership
dim()	dimension
$\Gamma()$	Gamma function
i	complex number
$\mathcal{H}$	hypothesis

ln	logarithm
$\mathcal{N}$	normal distribution
$rk()$	rank of a matrix
$sp()$	space
$\mathrm{tr}()$	trace of a matrix
vec	column-stacking operator
$vech$	column-stacking operator main diagonal and below a matrix

References

Akaike, H. [1981], 'Likelihood of a model and information criteria', *Journal of Econometrics* **16**, 3–14.

Amisano, G. and Giannini, C. [1997], *Topics in Structural VAR Econometrics*, 2nd edn, Springer, Berlin.

Anis, A. A. and Lloyd, E. H. [1976], 'The expected value of the adjusted rescaled Hurst range of independent normal summands', *Biometrika* **63**, 111–116.

Aydogan, K. and Booth, G. G. [1988], 'Are there long cycles in common stock returns?', *Southern Economic Journal* **55**, 141–149.

Baillie, R. T. [1996], 'Long memory processes and fractional integration in econometrics', *Journal of Econometrics* **73**, 5–59.

Barbosa, S. M. [2007], *mAr: Multivariate AutoRegressive analysis*.
URL: *http://CRAN.R-project.org*

Bera, A. K. and Jarque, C. M. [1980], 'Efficient tests for normality, heteroscedasticity, and serial independence of regression residuals', *Economic Letters* **6**(3), 255–259.

Bera, A. K. and Jarque, C. M. [1981], An efficient large-sample test for normality of observations and regression residuals, Working Papers in Econometrics 40, Australian National University, Canberra.

Beveridge, S. and Nelson, C. R. [1981], 'A new approach to decomposition of economic time series into permanent and transitory components with particular attention to measurement of the "business cycle"', *Journal of Monetary Economics* **7**, 151–174.

Bloomfield, P. [2000], *Fourier Analysis of Time Series: An Introduction*, 2nd edn, John Wiley and Sons, New York.

Box, G. E. P. and Jenkins, G. M. [1976], *Time Series Analysis: Forecasting and Control*, revised edn, Holden-Day, San Francisco.

Breitung, J., Brüggemann, R. and Lütkepohl, H. [2004], Structural vector autoregressive modeling and impulse responses, *in* H. Lütkepohl and M. Krätzig, eds, 'Applied Time Series Econometrics', Cambridge University Press, Cambridge, chapter 4, pp. 159–196.

Breusch, T. [1978], 'Testing for autocorrelation in dynamic linear models', *Australian Economic Papers* **17**, 334–355.

Britton, E., Fisher, P. and Whitley, J. [1998], 'The inflation report projections: understanding the fan chart', *Bank of England Quarterly Bulletin* **38**, 30–37.

Campbell, J. Y. and Perron, P. [1991], Pitfalls and opportunities: What macroeconomists should know about unit roots, *in* 'NBER Macroeconomic Annual 1991', National Bureau of Economic Research, Cambridge, MA, pp. 141–218.

Canova, F. and Hansen, B. E. [1995], 'Are seasonal patterns constant over time? A test for seasonal stationarity', *Journal of Business and Economic Statistics* **13**, 237–252.

Chambers, J. M. [2000], *Programming with Data: A Guide to the S Language*, 3rd edn, Springer-Verlag, New York.

Dalgaard, P. [2002], *Introductory Statistics with R*, Springer-Verlag, New York.

Davidson, R. and MacKinnon, J. G. [1993], *Estimation and Inference in Econometrics*, Oxford University Press, London.

Davies, R. B. and Harte, D. S. [1987], 'Tests for Hurst effects', *Biometrika* **74**, 95–102.

Dickey, D. A. and Fuller, W. A. [1979], 'Distributions of the estimators for autoregressive time series with a unit root', *Journal of the American Statistical Association* **74**, 427–431.

Dickey, D. A. and Fuller, W. A. [1981], 'Likelihood ratio statistics for autoregressive time series with a unit root', *Econometrica* **49**, 1057–1072.

Dickey, D. A., Hasza, D. P. and Fuller, W. A. [1984], 'Testing for unit roots in seasonal time series', *Journal of the American Statistical Association* **5**, 355–367.

Dickey, D. A. and Pantula, S. G. [1987], 'Determining the order of differencing in autoregressive process', *Journal of Business & Economic Statistics* **5**(4), 455–461.

Diebold, F. X. and Rudebusch, G. D. [1989], 'Long memory and persistence in aggregate output', *Journal of Monetary Economics* **24**, 189–209.

Durbin, J. and Watson, G. S. [1950], 'Testing for serial correlation in least-squares regression I', *Biometrika* **37**, 409–428.

Durbin, J. and Watson, G. S. [1951], 'Testing for serial correlation in least-squares regression II', *Biometrika* **38**, 159–178.

Durbin, J. and Watson, G. S. [1971], 'Testing for serial correlation in least-squares regression III', *Biometrika* **58**, 1–19.

Edgerton, D. and Shukur, G. [1999], 'Testing autocorrelation in a system perspective', *Econometric Reviews* **18**, 343–386.

Efron, B. and Tibshirani, R. [1993], *An Introduction to the Bootstrap*, Chapman & Hall, New York.

Elliott, G., Rothenberg, T. J. and Stock, J. H. [1996], 'Efficient tests for an autoregressive unit root', *Econometrica* **64**(4), 813–836.

Engle, R. [1982], 'Autoregressive conditional heteroscedasticity with estimates of the variance of United Kingdom inflation', *Econometrica* **50**, 987–1007.

Engle, R. F. and Granger, C. W. J. [1987], 'Co-integration and error correction: Representation, estimation, and testing', *Econometrica* **55**(2), 251–276.

Engle, R. F., Granger, C. W. J. and Hallman, J. [1988], 'Merging short- and long-run forecasts: An application of seasonal co-integration to monthly electricity sales forecasting', *Journal of Econometrics* **40**, 45–62.

Engle, R. and Yoo, S. [1987], 'Forecasting and testing in cointegrated systems', *Journal of Econometrics* **35**, 143–159.

Fox, J. [2004], *Rcmdr: R Commander*.
    URL: *http://CRAN.R-project.org*

Fox, J. [2007], *car: Companion to Applied Regression*.
   URL: *http://CRAN.R-project.org*

Fraley, C., Leisch, F. and Maechler, M. [2004], *fracdiff: Fractionally Differenced ARIMA (p,d,q) Models*.
   URL: *http://CRAN.R-project.org*

Franses, P. H. and Hobijn, B. [1997], 'Critical values for unit root tests in seasonal time series', *Journal of Applied Statistics* **24**, 25–48.

Frisch, R. and Waugh, F. V. [1933], 'Partial time regressions as compared with individual trends', *Econometrica* **1**, 387–401.

Fuller, W. A. [1976], *Introduction to Statistical Time Series*, John Wiley & Sons, New York.

Geweke, J. and Porter-Hudak, S. [1983], 'The estimation and application of long memory time series', *Journal of Time Series Analysis* **4**, 221–238.

Gilbert, P. [1993], State space and ARMA models: An overview of the equivalence, Working Paper 93-4, Bank of Canada, Ottawa, Canada.
   URL: *http:www.bank-banque-canada.ca/pgilbert/*

Gilbert, P. [1995], 'Combining var estimation and state space model reduction for simple good predictions', *Journal of Forecasting: Special Issue on VAR Modelling* **14**, 229–250.

Gilbert, P. [2000], 'A note on the computation of time series model roots', *Applied Economics Letters* **7**, 423–424.

Gilbert, P. [2004], *dse1: Dynamic Systems Estimation (time series package)*.
   URL: *http://www.bank-banque-canada.ca/pgilbert*

Godfrey, L. [1978], 'Testing for higher order serial correlation in regression equations when the regressors include lagged dependent variables', *Econometrica* **46**, 1303–1310.

Granger, C. [1969], 'Investigating causal relations by econometric models and cross-spectral methods', *Econometrica* **37**, 424–438.

Granger, C. W. J. [1980], 'Long memory relationships and the aggregation of dynamic models', *Journal of Econometrics* **14**, 227–238.

Granger, C. W. J. [1981], 'Some properties of time series data and their use in econometric model specification', *Journal of Econometrics* **16**, 150–161.

Granger, C. W. J. and Joyeux, R. [1980], 'An introduction to long-memory time series models and fractional differencing', *Journal of Time Series Analysis* **1**, 15–29.

Granger, C. W. J. and Newbold, P. [1974], 'Spurious regressions in econometrics', *Journal of Econometrics* **2**, 111–120.

Grothendieck, G. [2005], *dyn: Time Series Regression*.
   URL: *http://CRAN.R-project.org*

Grothendieck, G. and Petzoldt, T. [2004], 'R help desk: Date and time classes in R', *R News* **4**(1), 29–32.
   URL: *http://CRAN.R-project.org/doc/Rnews/*

Hamilton, J. D. [1994], *Time Series Analysis*, Princeton University Press, Princeton, NJ.

Hannan, E. and Quinn, B. [1979], 'The determination of the order of an autoregression', *Journal of the Royal Statistical Society* **B41**, 190–195.

Haslett, J. and Raftery, A. E. [1989], 'Space-time modelling with long-memory dependence: Assessing Ireland's wind power resource (with Discussion)', *Applied Statistics* **38**, 1–50.

Hasza, D. P. and Fuller, W. A. [1982], 'Testing for nonstationary parameter specifications in seasonal time series models', *The Annals of Statistics* **10**, 1209–1216.

Hendry, D. F. [1980], 'Econometrics: Alchemy or science?', *Economica* **47**, 387–406.

Hendry, D. F. [1986], 'Econometric modelling with cointegrated variables: An overview', *Oxford Bulletin of Economics and Statistics* **48**(3), 201–212.

Hendry, D. F. [2004], 'The Nobel Memorial Prize for Clive W. J. Granger', *Scandinavian Journal of Economics* **106**(2), 187–213.

Hendry, D. F. and Anderson, G. J. [1977], Testing dynamic specification in small simultaneous systems: An application to a model of building society behaviour in the United Kingdom, *in* M. D. Intriligator, ed., 'Frontiers in Quantitative Economics', Vol. 3, North-Holland, Amsterdam, pp. 361–383.

Holden, D. and Perman, R. [1994], Unit roots and cointegration for the economist, *in* B. B. Rao, ed., 'Cointegration for the Applied Economist', The MacMillan Press Ltd., London, chapter 3.

Hooker, R. H. [1901], 'Correlation of the marriage-rate with trade', *Journal of the Royal Statistical Society* **64**, 485–492.

Hosking, J. R. M. [1981], 'Fractional differencing', *Biometrika* **68**(1), 165–176.

Hurst, H. [1951], 'Long term storage capacity of reservoirs', *Transactions of the American Society of Civil Engineers* **116**, 770–799.

Hylleberg, S., Engle, R. F., Granger, C. W. J. and Yoo, B. S. [1990], 'Seasonal integration and cointegration', *Journal of Econometrics* **69**, 215–238.

Hyndman, R. J. [2007], *forecast: Forecasting functions for time series*.
  URL: *http://CRAN.R-project.org*

Jaconson, T., Vredin, A. and Warne, A. [1997], 'Common trends and hysteresis in scandinavian unemployment', *European Economic Review* **41**, 1781–1816.

Jarque, C. M. and Bera, A. K. [1987], 'A test for normality of observations and regression residuals', *International Statistical Review* **55**, 163–172.

Johansen, S. [1988], 'Statistical analysis of cointegration vectors', *Journal of Economic Dynamics and Control* **12**, 231–254.

Johansen, S. [1991], 'Estimation and hypothesis testing of cointegration vectors in Gaussian vector autoregressive models', *Econometrica* **59**(6), 1551–1580.

Johansen, S. [1995], *Likelihood-Based Inference in Cointegrated Vector Autoregressive Models*, Advanced Texts in Econometrics, Oxford University Press, Oxford.

Johansen, S. and Juselius, K. [1990], 'Maximum likelihood estimation and inference on cointegration — with applications to the demand for money', *Oxford Bulletin of Economics and Statistics* **52**(2), 169–210.

Johansen, S. and Juselius, K. [1992], 'Testing structural hypothesis in a multivariate cointegration analysis of the PPP and the UIP for UK', *Journal of Econometrics* **53**, 211–244.

Judge, G. G., Griffiths, W. E., Hill, R. C., Lütkepohl, H. and Lee, T. [1985], *The Theory and Practice of Econometrics*, 2nd edn, John Wiley and Sons, New York.

Kuan, C.-M. and Hornik, K. [1995], 'The generalized fluctuation test: A unifying view', *Econometric Reviews* **14**, 135–161.

Kwiatkowski, D., Phillips, P. C. B., Schmidt, P. and Shin, Y. [1992], 'Testing the null hypothesis of stationarity against the alternative of a unit root: How sure are we that economic time series have a unit root?', *Journal of Econometrics* **54**, 159–178.

Ljung, G. M. and Box, G. E. P. [1978], 'On a measure of lack of fit in time series models', *Biometrika* **65**, 553–564.

Lo, A. W. [1991], 'Long-term memory in stock market prices', *Econometrica* **59**(5), 1279–1313.

López-de Lacalle, J. and Díaz-Emparanza, I. [2004], *uroot: Unit root tests and graphics for seasonal time series.*
URL: *http://CRAN.R-project.org*

Lütkepohl, H. [2006], *New Introduction to Multiple Time Series Analysis*, Springer, New York.

Lütkepohl, H. and Saikkonen, P. [2000], 'Testing for the cointegration rank of a VAR process with a time trend', *Journal of Econometrics* **95**, 177–198.

Lütkepohl, H., Saikkonen, P. and Trenkler, C. [2004], 'Testing for the cointegrating rank of a VAR with level shift at unknown time', *Econometrica* **72**(2), 647–662.

MacKinnon, J. [1991], Critical values for cointegration tests, *in* R. F. Engle and C. W. J. Granger, eds, 'Long-Run Economic Relationships: Readings in Cointegration', Advanced Texts in Econometrics, Oxford University Press, Oxford, UK, chapter 13.

Mandelbrot, B. B. [1972], 'Statistical methodology for non periodic cycles: From the covariance to R/S analysis', *Annals of Economic and Social Measurement* **1**, 259–290.

Mandelbrot, B. B. [1975], 'A fast fractional Gaussian noise generator', *Water Resources Research* **7**, 543–553.

Mandelbrot, B. B. and Wallis, J. [1968], 'Noah, Joseph and operational hydrology', *Water Resources Research* **4**, 909–918.

Mandelbrot, B. B. and Wallis, J. [1969], 'Robustness of the rescaled range R/S in the measurement of noncyclical long-run statistical dependence', *Water Resources Research* **5**, 967–988.

McLeod, A. I. and Hipel, K. W. [1978], 'Preservation of the rescaled adjusted range, 1: A reassessment of the Hurst phenomenon', *Water Resources Research* **14**, 491–508.

Mehra, Y. P. [1994], Wage growth and the inflation process: An empirical approach, *in* B. B. Rao, ed., 'Cointegration for the Applied Economist', The MacMillan Press Ltd., London, chapter 5.

Nelson, C. R. and Plosser, C. I. [1982], 'Trends and random walks in macroeconomic time series', *Journal of Monetary Economics* **10**, 139–162.

Newey, W. and West, K. [1987], 'A simple positive definite, heteroscedasticity and autocorrelation consistent covariance matrix', *Econometrica* **55**, 703–705.

Osborn, D. R., Chui, A. P. L., Smith, J. P. and Birchenhall, C. R. [1988], 'Seasonality and the order of integration for consumption', *Oxford Bulletin of Economics and Statistics* **54**, 361–377.

Osterwald-Lenum, M. [1992], 'A note with quantiles of the asymptotic distribution of the maximum likelihood cointegration rank test statistic', *Oxford Bulletin of Economics and Statistics* **54**(3), 461–472.

Ouliaris, S., Park, J. Y. and Phillips, P. C. B. [1989], Testing for a unit root in the presence of a maintained trend, *in* B. Raj, ed., 'Advances in Econometrics and Modelling', Kluwer Academic Publishers, Dordrecht, pp. 7–28.

Perron, P. [1988], 'Trends and random walks in macroeconomic time series', *Journal of Economic Dynamics and Control* **12**, 297–332.

Perron, P. [1989], 'The Great Crash, the Oil Price Shock, and the Unit Root Hypothesis', *Econometrica* **57**(6), 1361–1401.

Perron, P. [1990], 'Testing for a unit root in a time series with a changing mean', *Journal of Business & Economic Statistics* **8**(2), 153–162.

Perron, P. [1993], 'Erratum: The Great Crash, the Oil Price Shock and the Unit Root Hypothesis', *Econometrica* **61**(1), 248–249.

Perron, P. and Vogelsang, T. J. [1992], 'Testing for a unit root in a time series with a changing mean: Corrections and extensions', *Journal of Business & Economic Statistics* **10**, 467–470.

Phillips, P. C. B. [1986], 'Understanding spurious regressions in econometrics', Cowles Foundation for Research in Economics, Yale University, Cowles Foundation Paper 667.

Phillips, P. C. B. and Ouliaris, S. [1990], 'Asymptotic properties of residual based tests for cointegration', *Econometrica* **58**, 165–193.

Phillips, P. C. B. and Perron, P. [1988], 'Testing for a unit root in time series regression', *Biometrika* **75**(2), 335–346.

Quinn, B. [1980], 'Order determination for multivariate autoregression', *Journal of the Royal Statistical Society* **B42**, 182–185.

R Development Core Team [2008], *R: A Language and Environment for Statistical Computing*, R Foundation for Statistical Computing, Vienna, Austria. ISBN 3-900051-07-0.
URL: *http://www.R-project.org*

Ripley, B. D. and Hornik, K. [2001], 'Date-time classes', *R News* **1**(2), 8–11.
URL: *http://CRAN.R-project.org/doc/Rnews/*

Robinson, P. M. [1994], 'Efficient tests of nonstationary hypotheses', *Journal of the American Statistical Association* **89**(428), 1420–1437.

Sargan, J. D. [1964], Wages and prices in the United Kingdom: A study in econometric methodology (with Discussion), *in* P. E. Hart, G. Mills and J. K. Whitaker, eds, 'Econometric Analysis for National Economic Planning', Vol. 16 of *Colsion Papers*, Butterworth Co., London, pp. 25–63.

Sargan, J. D. and Bhargava, A. [1983], 'Testing residuals from least squares regression for being generated by the Gaussian random walk', *Econometrica* **51**, 153–174.

Schmidt, P. and Phillips, P. C. B. [1992], 'LM tests for a unit root in the presence of deterministic trends', *Oxford Bulletin of Economics and Statistics* **54**(3), 257–287.

Schotman, P. and van Dijk, H. K. [1991], 'On Bayesian routes to unit roots', *Journal of Applied Econometrics* **6**, 387–401.

Schwarz, H. [1978], 'Estimating the dimension of a model', *The Annals of Statistics* **6**, 461–464.

Shapiro, S. S. and Wilk, M. B. [1965], 'An analysis of variance test for normality (complete samples)', *Biometrika* **52**, 591–611.

Shapiro, S. S., Wilk, M. B. and Chen, H. J. [1968], 'A comparative study of various tests of normality', *Journal of the American Statistical Association* **63**, 1343–1372.

Sims, C. [1980], 'Macroeconomics and reality', *Econometrica* **48**, 1–48.

Sowell, F. B. [1992], 'Modeling long run behaviour with fractional ARIMA model', *Journal of Monetary Economics* **29**, 277–302.

Spanos, A. [1986], *Statistical Foundations of Econometric Modelling*, Cambridge University Press, Cambridge.

Stock, J. H. [1987], 'Asymptotic properties of least squares estimators of cointegrating vectors', *Econometrica* **55**, 1035–1056.

Trapletti, A. and Hornik, K. [2004], *tseries: Time Series Analysis and Computational Finance*.
URL: *http://CRAN.R-project.org*

Trenkler, C. [2003], 'A new set of critical values for systems cointegration tests with a prior adjustment for deterministic terms', *Economics Bulletin* **3**(11), 1–9.

Venables, W. N. and Ripley, B. D. [2002], *Modern Applied Statistics with S*, 4th edn, Springer-Verlag, New York.

Wallis, K. F. [1974], 'Seasonal adjustment and relations between variables', *Journal of the American Statistical Association* **69**, 18–31.

White, H. [1984], *Asymptotic Theory for Econometricians*, Academic Press, New York.

Würtz, D. [2007a], *fArma: Rmetrics - ARMA Time Series Modelling*.
URL: *http://CRAN.R-project.org*

Würtz, D. [2007b], *fUnitRoots: Rmetrics - Trends and Unit Roots*.
URL: *http://CRAN.R-project.org*

Yule, G. U. [1926], 'Why do we sometimes get nonsense-correlations between time series? A study in sampling and the nature of time series', *Journal of the Royal Statistical Society* **89**, 1–64.

Zeileis, A. [2006], *dynlm: Dynamic Linear Regression*.
URL: *http://CRAN.R-project.org*

Zeileis, A. and Hothorn, T. [2002], 'Diagnostic checking in regression relationships', *R News* **2**(3), 7–10.
URL: *http://CRAN.R-project.org*

Zeileis, A., Leisch, F., Hornik, K. and Kleiber, C. [2005], 'Monitoring structural change in dynamic econometric models', *Journal of Applied Econometrics* **20**(1), 99–121.

Zivot, E. and Andrews, D. W. K. [1992], 'Further evidence on the Great Crash, the Oil-Price Shock, and the Unit-Root Hypothesis', *Journal of Business & Economic Statistics* **10**(3), 251–270.

Name Index

Akaike, H. 15, 25, 61, 165
Amisano, G. 43, 44
Anderson, G. 75
Andrews, D. 110, 111, 121, 166
Anis, A. 68
Aydogan, K. 68

Baillie, R. 62
Barbosa, S. 26
Bera, A. 16, 31, 165
Beveridge, S. 54
Bhargava, A. 76
Birchenhall, C. 59, 115, 118
Bloomfield, P. 64
Booth, G. 68
Box, G. 15, 16, 61, 165
Brüggemann, R, 146, 149
Breitung, J. 146, 149
Breusch, T. 29
Britton, E. 37

Campbell, J. 53, 70, 79
Canova, F. 116
Chambers, J. 86
Chen, H. 16

Chui, A. 59, 115, 118

Dalgaard, P. ix
Davidson, R. 5, 149
Davies, R. 68
Dickey, D. 59–62, 92–94, 96, 113,
        149, 165
Diebold, F. 70
Durbin, J. 74

Edgerton, D. 29
Efron, B. 40
Elliott, G. 98, 99, 105, 165
Engle, R. 23, 30, 57, 59, 75, 76, 86,
        115, 116, 122, 165

Fisher, P. 37
Fox, J. 75, 82
Franses, P. 113, 117
Frisch, R. 74
Fuller, W. 59–61, 71, 92–94, 96,
        113, 149, 165

Geweke, J. 69, 71
Giannini, C. 43, 44

Gilbert, P.   26
Godfrey, L.   29
Granger, C.   ix, 23, 34, 57, 59, 62,
    65, 74–76, 115, 116, 165
Griffiths, W.   19, 64
Grothendieck, G.   78, 162

Hallmann, J.   59
Hamilton, J.   5, 9, 10, 19, 30, 93
Hannan, E.   15, 25
Hansen, B.   116
Harte, D.   68
Haslett, J.   70
Hasza, D.   59, 113
Hendry, D.   ix, 74, 75
Hill, R.   19, 64
Hipel, K.   62
Hobijn, B.   113, 117
Holden, D.   91, 121
Hooker, R.   74
Hornik, K.   16, 33, 162
Hosking, J.   65
Hurst, H.   66–68
Hylleberg, S.   59, 115, 116, 165
Hyndman, R.   18

Jacobson, T.   146
Jarque, C.   16, 31, 165
Jenkins, G.   15, 16
Johansen, S.   81, 129, 130,
    135–142, 144
Joyeux, R.   65
Judge, G.   19, 64
Juselius, K.   81, 129, 130,
    135–142, 144

Kleiber, C.   33
Kuan, C.   33
Kwiatkowski, D.   103, 105, 165

Lütkepohl, H.   19, 24, 31, 35, 64,
    143, 144, 146, 149
Lee, T.   64
Leisch, F.   33
Ljung, G.   16, 61, 165

Lloyd, E.   68
Lo, A.   68, 71

MacKinnon, J.   5, 60, 76, 86, 93,
    122, 149
Mandelbrot, B.   68
McLeod, A.   62
Mehra, Y.   127, 159

Nelson, C.   3, 54, 99, 101, 104, 105,
    110, 111
Newbold, P.   74
Newey, W.   69

Osborn, D.   59, 115, 118
Osterald-Lenum, M.   132
Ouliaris, S.   60, 76, 86, 121, 124, 125

Pantula, S.   62
Park, J.   60
Perman, R.   91, 121
Perron, P.   53, 70, 79, 95, 108, 110,
    111, 118, 121, 166
Petzoldt, T.   162
Phillips, P.   60, 74, 76, 86, 95,
    100, 103, 105, 121, 124, 125,
    165, 166
Plosser, C.   3, 99, 101, 104, 105,
    110, 111
Porter-Hudak, S.   69, 71

Quinn, B.   15, 25

Raftery, A.   70
Ripley, B.   ix, 64, 162
Robinson, P.   65
Rothenberg, T.   98, 99, 105, 165
Rudebusch, G.   70

Saikkonen, P.   143, 144
Sargan, J.   75, 76
Schmidt, P.   100, 103, 105, 165, 166
Schotman, P.   3
Schwarz, H.   15, 25, 61, 166
Shapiro, S.   16

Shin, Y.   103, 105, 165
Shukur, G.   29
Sims, C.   23
Smith, J.   59, 115, 118
Sowell, F.   70
Spanos, A.   5
Stock, J.   76, 98, 99, 105, 165

Tibshirani, R.   40
Trapletti, A.   16
Trenkler, C.   143, 144

van Dijk, H.   3
Venables, W.   ix, 64
Vogelsang, T.   108
Vredin, A.   146

Würtz, D.   7, 76
Wallis, J.   68
Wallis, K.   112
Warne, A.   146
Watson, G.   74
Waugh, F.   74
West, K.   69
White, H.   5
Whitley, J.   37
Wilk, M.   16

Yoo, B.   59, 76, 86, 115, 116, 122, 165
Yule, G.   74

Zeileis, A.   33, 75, 78
Zivot, E.   110, 111, 121, 166

Function Index

%*% 133

abline() 12, 20
ablrtest() 139
acf() 7, 17, 66
ADF.test() 91, 116
adf.test() 91
adftest() 91
alrtest() 135, 136
arch.test() 30, 31, 85, 150
args() 30, 38, 40, 42, 46, 48, 131
arima() 10, 12, 16, 17
arima.sim() 7, 10, 66, 83
ARMA() 26, 28, 46, 47
armaFit() 7
armaRoots() 7
armaSim() 7
armaTrueacf() 7
array() 28, 46, 47
as.numeric() 17
attach() 92, 96, 122, 126, 131
auto.arima() 17, 18
axis() 56

bh5lrtest() 140

bh6lrtest() 142
blrtest() 137, 138
Box.test() 16, 17

ca.jo() 82, 83, 131, 144, 151–155
ca.po() 125, 126
cajolst() 144
cajorls() 83, 84, 154
causality() 36
cbind() 77, 126, 131, 133, 154
CH.rectest() 116
CH.test() 116
class() 30, 38, 48, 83, 85
coefficients() 154
colnames() 28, 46, 47, 77, 123
crossprod() 154
cumsum() 56, 68, 74, 77, 83, 108

data() 17, 92, 96, 99, 101, 104, 111, 117, 122, 126, 131, 144, 147, 161
data.frame() 77, 83
diag() 28, 46, 47, 142, 154
diff() 77, 92, 96, 99, 123, 148
durbin.watson() 75
dwtest() 74

dyn()   78
dynlm()   78

efp()   33
embed()   77, 123
example()   131, 139

fanchart()   37, 38
fevd()   42, 48, 85, 158
fevd.svarest()   48
fevd.varest()   42
filter()   10, 12
forecast()   19
fracdiff()   70
fracdiff.sim()   66, 68, 69

HEGY.test()   116, 117

Im()   10, 12
irf()   40, 48, 85, 157
irf.svarest()   48
irf.varest()   40
irts()   162
its()   162

jarque.bera.test()   16, 122

kpss.test()   103
kronecker()   154

lag()   123
latex()   162
layout()   7, 17, 66
legend()   12, 56
library()   17, 28, 46, 47, 66, 68,
    69, 74, 92, 111, 117, 122,
    126, 131, 147, 161
lines()   20, 56
lm()   53, 69, 74, 77, 122, 123
log()   68, 69, 99, 104, 111
logLik()   16, 17
lttest()   130

matrix()   7, 17, 28, 46, 47, 66, 155
max()   68
methods()   30, 48, 85

min()   68
Mod()   10, 12

na.omit()   17, 99, 101, 104, 111
names()   85, 131
normality.test()   30, 32, 85, 150
nrow()   154

optim()   44, 45

pacf()   7, 17
par()   7, 17, 56, 66, 147
pchisq()   17
plot()   12, 17, 20, 92, 111, 147,
    150, 157
plot.forecast()   19
plot.ts()   7, 28, 56, 66
plot.varcheck()   30, 32
plot.varfevd()   42, 48
plot.varirf()   40, 48
plot.varprd()   37, 38
points()   12
polyroot()   10, 12
pp.test()   96
predict()   7, 20, 38, 85
predict.varest()   37, 38

qqnorm()   16

rbind()   142
Re()   10, 12
rep()   108
residuals()   17, 53, 77, 122, 154
rnorm()   12, 56, 74, 77, 83, 108
roots()   26, 28

sd()   68
seq()   12, 53
serial.test()   30, 85, 150
set.seed()   7, 56, 66, 68, 69, 74,
    77, 83, 108
shapiro.test()   16, 17
simulate()   26, 28, 46, 47
sin()   69
slotNames()   83, 85

solve()   125, 133, 154
spectrum()   64, 66, 69
sqrt()   12, 69, 154
stability()   33, 34, 85, 150
str()   161
summary()   69, 83, 122, 126, 131, 136,
      138, 140, 142, 144, 147, 148, 152
SVAR()   44–47
SVEC()   155, 156

time()   161
timeSeries()   162
ts()   17, 20, 92, 96, 117, 122, 123,
      126, 133, 161
tsdiag()   16, 17
tsp()   161

unitrootTable()   76
update()   156
ur.df()   91–93, 122, 148, 149
ur.ers()   99
ur.kpss()   103, 104
ur.pp()   96
ur.sp()   101
ur.za()   111
urkpssTest()   103
urootgui()   116
urppTest()   96

VAR()   25, 26, 28, 46, 47, 150
VARselect()   25, 26, 28, 149
vec2var()   85

window()   122, 123, 126, 161

zoo()   162

Subject Index

ADF *see* unit root test
AIC *see* Akaike information criteria
Akaike information criteria 25, 61, 165
AR *see* autoregressive process
ARFIMA *see* autoregressive fractional integrated moving average process
ARIMA *see* autoregressive integrated moving average process
ARMA *see* autoregressive moving average process
autoregressive fractional integrated moving average process 62, 65, 70, 165
autoregressive integrated moving average process 54, 165
autoregressive moving average process 14, 165
  characteristic polynomial 15
  stationarity 15
autoregressive process 6, 165
  AR(p) 7
  autocorrelation 7

characteristic polynomial 8
first-order 6
partial autocorrelation 7
stability 10

Bayesian information criteria 25, 61, 99, 165
Beveridge-Nelson decomposition 54
BIC *see* Bayesian information criteria
Box-Jenkins method 15

CI *see* cointegration
classic linear regression model 73
cointegrating Durbin-Watson statistic 76, 165
cointegration 75, 165
  Engle-Granger 76, 121
  Phillips-Ouliaris 124
  representation theorem 77
  superconsistency 76
  two-step 76
CRDW *see* cointegrating Durbin-Watson statistic

data-generating process   3, 54, 55, 165
   cyclical trend   54
   deterministic trend   54
   stochastic trend   54
DF   *see* unit root test
DGP   *see* data-generating process
Durbin-Watson statistic   74, 76
DW   *see* Durbin-Watson statistic

ECM   *see* error-correction model
eigenvalue   24
ergodicity   5, 11
error-correction model   77, 165
ERS   *see* unit root test

fan chart   37
FEVD   *see* forecast error variance
       decomposition
final prediction error   25
fractional integration
   R/S statistic   66
   anti-persistent   65
   autocorrelation   65
   Geweke and Porter-Hudak   69
   Hurst coefficient   67
   intermediate memory   65
   long-memory   62, 65
   spectral density   64

Granger causality   77, 123

Hannan and Quinn information
       criteria   25
HEGY   *see* unit root test
HQ   *see* Hannan and Quinn
       information criteria

I(d)   *see* integrated of order d
i.d.   *see* independently distributed
i.i.d.   *see* independently and
       identically distributed
impulse response analysis   38, 156
independently and identically
       distributed   9, 53, 165
independently distributed   165
information criteria   25

integrated of order d 57, 165
integration   57
   fractional   62
   seasonal   57, 59, 166
IRA   *see* impulse response
       analysis

Jarque-Bera test   16

KPSS   *see* unit root test

Ljung-Box test   16, 61
Long-memory   *see* fractional
       integration

MA   *see* moving average process
moving average process   10, 54, 166
   autocorrelation   12
   MA(q)   11
   partial autocorrelation   12
   stability   13

near integrated   166
NI   *see* near integrated
nonsense regression   74

OLS   *see* ordinary least-squares
ordinary least-squares   8, 76, 166

Portmanteau test   *see* Ljung-Box
       test
PP   *see* unit root test
purchasing power parity (PPP)   129

quantiles plot   16

random walk   55
   drift   55
   pure   55

S3 class   85, 162
S4 class   85, 162
sample size   166
SC   *see* Schwarz criteria
Schwarz criteria   25, 61, 166
Shapiro-Wilk test   16
SI   *see* integration

SP   *see* unit root test
spectral density   66
spurious regression   74
stationarity   3
  covariance   4
  difference   53
  second-order   4
  strict   4
  trend   53
  weak   3
stochastic process   3
stochastic seasonality   58
structural vector autoregressive
    model   23, 43
  A-model   45
  B-model   46
  forecast error variance
      decomposition   48
  impulse response function   48
  likelihood   44
  overidentification   45
  scoring algorithm   46
structural vector error-correction
    model   23, 145, 146
SVAR   *see* structural vector
    autoregressive model
SVEC   *see* structural vector error
    correction

T   *see* sample size
time series   3
  decomposition   53
  non-stationary   55
ts   *see* time series

uncovered interest rate parity (UIP)
    129
unit circle   10, 53
unit root   6, 54, 76
  process   55
  seasonal   59, 112
  structural break   107
unit root test
  augmented Dickey-Fuller   60, 76,
    91, 165
  bottom-up   62

  Canova-Hansen   116
  Dickey-Fuller   59, 76, 91
  Dickey-Hasza-Fuller   59, 113
  Elliott-Rothenberg-Stock   98,
    165
  general-to-specific   60
  Hylleberg-Engle-Granger-Yoo
    59, 115, 165
  Kwiatkowski-Phillips-Schmidt-
    Shin   103, 165
  Osborn   59, 115
  Perron   108
  Phillips-Perron   95, 166
  Schmidt-Phillips   100, 166
  testing procedure   61
  Zivot-Andrews   110, 166

VAR   *see* vector autoregressive
    model
var   *see* variance
variance   166
VECM   *see* vector error-correction
    model
vector autoregressive model   23,
    79, 166
  causality   35
  companion form   24
  definition   23
  fan chart   37
  forecast   36
  forecast error variance
      decomposition   41
  heteroscedasticity   28, 30
  impulse response function   38
  normality   28
  normality test   31
  serial correlation   28
  stability   24
  structural stability   29, 33
  Wold moving average   24
vector error-correction model   23,
    78, 79, 129, 166
  adjustment vectors   80
  canonical correlation   81
  cointegration vectors   80

estimator for adjustment vectors
    82
estimator for cointegration
    vectors   81
estimator for variance-covariance
    matrix   82
hypothesis on α 135, 136
hypothesis on α and β 139
hypothesis on β 136, 142
Johansen method   81
loading vectors   80
long-run form   79
maximal eigenvalue statistic   81
structural shift   143
trace statistic   81
transitory form   79
weak exogenity   134

white noise   6
  Gaussian   6
  independent   6
  normal   6
Wold representation   24

ZA   *see* unit root test

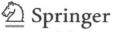

Time Series Analysis with Applications in R

Jonathan D. Cryer and Kung-Sik Chan

Time Series Analysis With Applications in R, Second Ed., presents an accessible approach to understanding time series models and their applications. Although the emphasis is on time domain ARIMA models and their analysis, the new edition devotes two chapters to the frequency domain and three to time series regression models, models for heteroscedasticty, and threshold models. All of the ideas and methods are illustrated with both real and simulated data sets. A unique feature of this edition is its integration with the R computing environment.

2008. 2nd Ed., 494 pp. (Springer Texts in Statistics) Hardcover
ISBN 0-387-75958-6

Time Series Analysis and Its Applications with R Examples

Robert H. Shumway and David S. Stoffer

This book presents a balanced and comprehensive treatment of both time and frequency domain methods with accompanying theory. Numerous examples using non-trivial data illustrate solutions to problems such as evaluating pain perception experiments using magnetic resonance imaging or monitoring a nuclear test ban treaty. It is designed to be useful as a graduate level text in the physical, biological and social sciences and as a graduate level text in statistics. Some parts may also serve as an undergraduate introductory courses the theory and methodology are separated to allow presentations on different levels.

2006. 2nd Ed. 575 pp. (Springer Texts in Statistics) Hardcover
ISBN 978-0-387-29317-2

Data Manipulation with R

Phil Spector

This book presents a wide array of methods applicable for reading data into R, and efficiently manipulating that data. In addition to the built-in functions, a number of readily available packages from CRAN (the Comprehensive R Archive Network) are also covered. All of the methods presented take advantage of the core features of R: vectorization, efficient use of subscripting, and the proper use of the varied functions in R that are provided for common data management tasks.

2008. 164 pp. (Use R) Softcover
ISBN 978-0-387-74730-9